Advanced

Modular

Mathematics

STATISTICS 3

Gerald Westover
Graham Smithers

**SECOND
EDITION**

COLLINS

Unit S3

Published by HarperCollins Publishers Limited
77–85 Fulham Palace Road
Hammersmith
London W6 8JB

www.CollinsEducation.com
On-line Support for Schools and Colleges

© National Extension College Trust Ltd 2000
First published 2000
ISBN 000 322523 2

This book was written by Gerald Westover and Graham Smithers for the National Extension College Trust Ltd. Part of the material was originally written by Mik Wisnieski and Clifford Taylor.

British Library Cataloguing in Publication Data
A catalogue record for this publication is available from the British Library.

Original internal design: Derek Lee
Cover design and implementation: Terry Bambrook
Project editors: Hugh Hillyard-Parker and Margaret Levin
Page layout: Mary Bishop, Eric Coles
Printed and bound: Scotprint, Musselburgh

The authors and publishers thank Dave Wilkins for his comments on this book.

The National Extension College is an educational trust and a registered charity with a distinguished body of trustees. It is an independent, self-financing organisation.

Since it was established in 1963, NEC has pioneered the development of flexible learning for adults. NEC is actively developing innovative materials and systems for distance-learning options from basic skills and general education to degree and professional training.

For further details of NEC resources that support Advanced Modular Mathematics, and other NEC courses, contact NEC Customer Services:

National Extension College Trust Ltd
18 Brooklands Avenue
Cambridge CB2 2HN
Telephone 01223 316644, Fax 01223 313586
Email resources@nec.ac.uk, Home page www.nec.ac.uk

You might also like to visit:

www.fireandwater.com
The book lover's website

UNIT S3
Contents

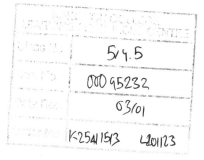

Advanced Modular Mathematics

FOREWORD   This book is one of a series covering the Edexcel Advanced Subsidiary (AS) and Advanced GCE in Mathematics. It covers all the subject material for Statistics 3 (Unit S3), examined from 2002 onwards.

While this series of text books has been structured to match the Edexcel specification, we hope that the informal style of the text and approach to important concepts will encourage other readers whose final exams are from other Boards to use the books for extra reading and practice. In particular, we have included references to the OCR syllabus (see below).

This book is meant to be *used*: read the text, study the worked examples and work through the Practice questions and Summary exercises, which will give you practice in the basic skills you need for maths at this level. Many exercises, and worked examples, are based on applications of the mathematics in this book. There are many books for advanced mathematics, which include many more exercises: use this book to direct your studies, making use of as many other resources as you can.

There are many features in this book that you will find particularly useful:

- Each **section** covers one discrete area of the new Edexcel specification. The order of topics is exactly the same as in the specification.

- **Practice questions** are given at regular intervals throughout each section. The questions are graded to help you build up your mathematical skills gradually through the section. The **Answers** to these questions come at the end of the relevant section.

- **Summary exercises** are given at the end of each section; these include more full-blown, exam-type questions. Full, worked solutions are given in a separate **Solutions** section at the end of the book.

- In addition, we have provided a complete **Practice examination paper**, which you can use as a 'dummy run' of the actual exam when you reach the end of your studies on S3.

- Alongside most of the headings in this book you will see boxed references, e.g. OCR **S3** 5.13.1 (a) These are for students following the OCR specification and indicate which part of that specification the topic covers.

- Your work on this book will provide opportunities for gathering evidence towards Key Skills, especially when you come to tackle your statistics project (see Section 6). Key Skills opportunities are indicated by a 'key' icon, for example: **C** 3.2 (see Appendix 5, p. 98, for more information).

The National Extension College has more experience of flexible-learning materials than any other body (see p. ii). This series is a distillation of that experience: Advanced Modular Mathematics helps to put you in control of your own learning.

1

Combinations of random variables

INTRODUCTION We have already seen the importance of the normal distribution but, as yet, we haven't considered a linear combination of two (or more) such distributions. For example, if the weights of empty tins are normally distributed with mean 30 g and standard deviation 0.7 g and the weights of their contents are independently normally distributed with a mean of 320 g and a standard deviation of 2.4 g, what will be the distribution of a full tin? It can be shown that a full tin will be normally distributed with mean 350 g (obvious?) and standard deviation 2.5 g (not so obvious?). That's the sort of problem we'll be studying in this section.

To begin with, though, we'll need to develop some rules and, for this, we'll be building on the work of section 7 in Unit S1.

Combinations of random variables OCR S3 5.13.2 (a)(i),(ii),(iii)

In this section we investigate the properties of linear combinations of independent random variables. The results we establish will apply to both discrete and continuous random variables but we will work with the simpler of the two – namely discrete random variables – in order to develop the relevant formulae more readily.

Example Let X be defined by the distribution

x	0	1	2
$P(X = x)$	$\frac{1}{2}$	$\frac{1}{3}$	$\frac{1}{6}$

and let Y be defined by the distribution

y	−1	0	1
$P(Y = y)$	$\frac{1}{4}$	$\frac{1}{4}$	$\frac{1}{2}$

Then we can form many new distributions from these two. Probably the simplest are $Z = X + Y$ and $W = X - Y$.

We do this by first working out the possible outcomes for Z and W and then the respective probabilities for these outcomes.

Working systematically Z and W can take values:

X	Y	$Z = X + Y$	$W = X - Y$	Probability
0	−1	−1	1	$\frac{1}{2} \times \frac{1}{4}$
0	0	0	0	$\frac{1}{2} \times \frac{1}{4}$
0	1	1	−1	$\frac{1}{2} \times \frac{1}{2}$
1	−1	0	2	$\frac{1}{3} \times \frac{1}{4}$
1	0	1	1	$\frac{1}{3} \times \frac{1}{4}$
1	1	2	0	$\frac{1}{3} \times \frac{1}{2}$
2	−1	1	3	$\frac{1}{6} \times \frac{1}{4}$
2	0	2	2	$\frac{1}{6} \times \frac{1}{4}$
2	1	3	1	$\frac{1}{6} \times \frac{1}{2}$

We shall deal first with Z. Tidying up gives:

z	−1	0	1	2	3
$P(Z = z)$	$\frac{3}{24}$	$\frac{5}{24}$	$\frac{9}{24}$	$\frac{5}{24}$	$\frac{2}{24}$

where for example the probability for $Z = 0$ has been obtained by evaluating $\frac{1}{2} \times \frac{1}{4} + \frac{1}{3} \times \frac{1}{4}$.

Note that Z is a proper random variable since the probabilities add up to 1. It's always worth checking this.

Example	Find $E(X)$, $E(Y)$, $E(Z)$ from the above example.

Solution	$E(X) = 0 \times \frac{1}{2} + 1 \times \frac{1}{3} + 2 \times \frac{1}{6} = \frac{2}{3}$

$E(Y) = -1 \times \frac{1}{4} + 0 \times \frac{1}{4} + 1 \times \frac{1}{2} = \frac{1}{4}$

$E(Z) = -1 \times \frac{3}{24} + 0 \times \frac{5}{24} + 1 \times \frac{9}{24} + 2 \times \frac{5}{24} + 3 \times \frac{2}{24} = \frac{11}{12}$

It can now be confirmed that $E(X + Y) = E(X) + E(Y)$

Example	Find $\text{Var}(X)$, $\text{Var}(Y)$, $\text{Var}(Z)$ from the same example.

Solution	We need $E(X^2)$, $E(Y^2)$, $E(Z^2)$

$E(X^2) = 0^2 \times \frac{1}{2} + 1^2 \times \frac{1}{3} + 2^2 \times \frac{1}{6} = 1$

$E(Y^2) = (-1)^2 \times \frac{1}{4} + 0^2 \times \frac{1}{4} + 1^2 \times \frac{1}{2} = \frac{3}{4}$

$E(Z^2) = (-1)^2 \times \frac{3}{24} + 0^2 \times \frac{5}{24} + 1^2 \times \frac{9}{24} + 2^2 \times \frac{5}{24} + 3^2 \times \frac{2}{24} = \frac{50}{24}$

and therefore

$$\text{Var}(X) = 1 - \left(\frac{2}{3}\right)^2 = \frac{5}{9}$$

$$\text{Var}(Y) = \frac{3}{4} - \left(\frac{1}{4}\right)^2 = \frac{11}{16}$$

$$\text{Var}(Z) = \frac{50}{24} - \left(\frac{11}{12}\right)^2 = \frac{179}{144}$$

It can now be seen that $\text{Var}(X) + \text{Var}(Y) = \text{Var}(X + Y)$

For W we have:

w	-1	0	1	2	3
$\text{P}(W = w)$	$\frac{6}{24}$	$\frac{7}{24}$	$\frac{7}{24}$	$\frac{3}{24}$	$\frac{1}{24}$

So $\text{E}(W) = \frac{1}{24}(-6 + 7 + 6 + 3) = \frac{10}{24} = \frac{5}{12}$

And $\text{E}(W^2) = \frac{1}{24}(6 + 7 + 12 + 9) = \frac{34}{24} = \frac{17}{12}$

Then $\text{Var}(W) = \frac{17}{12} - \left(\frac{5}{12}\right)^2 = \frac{179}{144}$

This too is equal to $\text{Var}(X) + \text{Var}(Y)$, *not* $\text{Var}(X) - \text{Var}(Y)$ as you might at first think.

The results observed in the previous example are not coincidence but are general to linear combinations of random variables.

The following properties are important.

If X and Y are random variables and a and b are any real numbers then:

$$\text{E}(aX + bY) = a\text{E}(X) + b\text{E}(Y)$$

$$\text{E}(aX - bY) = a\text{E}(X) - b\text{E}(Y)$$

and if X and Y are *independent* random variables and a and b are any real numbers then

$$\text{Var}(aX + bY) = a^2\,\text{Var}(X) + b^2\,\text{Var}(Y)$$

$$\text{Var}(aX - bY) = a^2\,\text{Var}(X) + b^2\,\text{Var}(Y)$$

The proofs of these properties are not expected at 'A' level, but the results are important and should be remembered. Note in particular the signs in the last of these.

Example

For random variable X, you are given $\text{E}(X) = 6$, $\text{Var}(X) = 5$

and for random variable Y, $\text{E}(Y) = 8$, $\text{Var}(Y) = 10$.

Find:

(a) $\text{E}(2X + 3Y)$

(b) $\text{Var}(2X + 3Y)$

(c) $\text{E}(3X - 4Y)$

(d) $\text{Var}(3X - 4Y)$

(e) $\text{E}(2X + Y + 5)$

(f) $\text{Var}(2X + Y + 5)$

Solution	(a) $E(2X + 3Y)$	$= 2E(X) + 3E(Y)$
		$= 2 \times 6 + 3 \times 8 = 35$
	(b) $Var(2X + 3Y)$	$= 4\,Var(X) + 9\,Var(Y)$
		$= 4 \times 5 + 9 \times 10 = 110$
	(c) $E(3X - 4Y)$	$= 3E(X) - 4E(Y)$
		$= 3 \times 6 - 4 \times 8 = -14$
	(d) $Var(3X - 4Y)$	$= 9\,Var(X) + 16\,Var(Y)$
		$= 9 \times 5 + 16 \times 10 = 205$ [Note the signs here.]
	(e) $E(2X + Y + 5)$	$= 2E(X) + E(Y) + E(5)$
		$= 2 \times 6 + 8 + 5 = 25$
	(f) $Var(2X + Y + 5)$	$= 4\,Var(X) + Var(Y) + Var(5)$
		$= 4 \times 5 + 10 + 0 = 30$
		Note that $Var(5) = 0$

Practice questions A

1 If X and Y are independent random variables and

$$E(X) = 30, \quad Var(X) = 4,$$
$$E(Y) = 20, \quad Var(Y) = 5,$$

find the mean and variance of:

(a) $X + Y$ (b) $X - Y$
(c) $2X + Y$ (d) $X - 2Y$
(e) $X + 20$ (f) $3X - 2Y$
(g) $X + Y + 40$ (h) $2X + 3Y + 50$

2 X and Y are independent random variables such that

$$E(X) = 40, \quad Var(X) = 5,$$
$$E(Y) = 34, \quad Var(Y) = 6.$$

Find the mean and variance of

(a) $X - 3$ (b) $X - 3Y$

3 (a) A single dice is thrown. Find the mean and variance of the score obtained.

(b) Three dice are thrown and the separate scores are denoted by the random variables D_1, D_2 and D_3.
Find the mean and variance of $D_1 + D_2 + D_3$, the total score obtained.

(c) The person throwing the dice always doubles the first score, adds 2 to the second score and subtracts 3 from the third score. What is the mean and variance of the total?

4 A single dice is thrown 20 times and the separate scores are added.

(a) What is the mean and variance of the final score obtained?

(b) What would be the mean and variance if you multiplied a single score by 20?

5 Rabbits have fleas with a mean of 48 and a variance of 5.

(a) If two rabbits are chosen at random and the total number of fleas is counted, what is the mean and variance of this total?

(b) If a single rabbit is randomly selected and the total number of fleas found is doubled, what is the mean and variance now?

6 A random variable X has a Poisson distribution with $E(X) = 16$.

(a) Write down the mean and variance of $5X$.

(b) Another random variable Y has a Poisson distribution with $E(Y) = 12$. Write down the mean and variance of $5X + 7Y + 8$. (Assume that X and Y are independent.)

7 In an examination, the mean mark was 56 with a variance of 4.

(a) If each mark is doubled, what is the new mean mark and what is its variance?

(b) If each mark is doubled and then 24 is subtracted, what is the mean mark now and what is its variance?

8 A random variable X has PDF

$$f(x) = \begin{cases} ax^2 & \text{for } 0 \le x \le 1 \\ 0 & \text{elsewhere} \end{cases}$$

where a is a constant.

(a) Find the value of a and hence the values of $E(X)$ and $\text{Var}(X)$.

(b) If random variables X_1 and X_2 both have the distribution above, find:

   (i) $E(X_1 + X_2)$ and $\text{Var}(X_1 + X_2)$

   (ii) $E(X_1 - X_2 + 5)$ and $\text{Var}(X_1 - X_2 + 5)$

Linear combinations of independent normal random variables

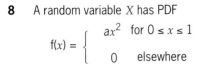

We'll now apply what we've learned so far to the combination of independent normal variables.

If X_1 and X_2 are independent normal variables with

$$X_1 \sim N(\mu_1 \; \sigma_1^2) \text{ and } X_2 \sim N(\mu_2, \sigma_2^2),$$

then our previous investigations lead us to conclude that:

- $X_1 + X_2 \sim N(\mu_1 + \mu_2, \sigma_1^2 + \sigma_2^2)$
- $X_1 - X_2 \sim N(\mu_1 - \mu_2, \sigma_1^2 + \sigma_2^2)$

  [Note the signs here.]

- $aX_1 + bX_2 \sim N(a\mu_1 + b\mu_2, a^2\sigma_1^2 + b^2\sigma_2^2)$

  (note that a and b are squared for the variance)

- $aX_1 - bX_2 \sim N(a\mu_1 - b\mu_2, a^2\sigma_1^2 + b^2\sigma_2^2)$

  [Once again, note the signs here.]

Let's try some examples

Example

The weight of fruit delivered into a can is normally distributed with mean 300 g and standard deviation 1.5 g. The weight of an empty can is normally distributed with mean 35 g and standard deviation 0.8 g. Calculate:

(a) the mean and standard deviation of the weight of a filled can

(b) the probability that a filled can will weigh more than 340 g.

Solution

In symbol form we have:

$X \sim N(300, 1.5^2)$ for the fruit and $Y \sim N(35, 0.8^2)$ for the can

(a) It follows that $X + Y \sim N(300 + 35, 1.5^2 + 0.8^2)$

    i.e. $X + Y \sim N(335, 2.89)$

    $\therefore$ Mean $= 335$ g and standard deviation $= \sqrt{2.89} = 1.7$ g

(b) Since the question asks for $P(X + Y > 340)$, we need to find the shaded area in Fig. 1.1:

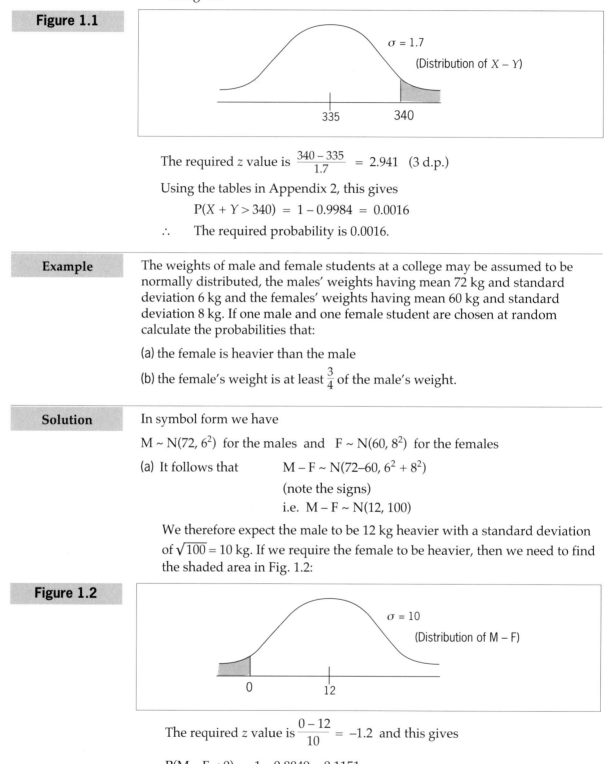

The required z value is $\dfrac{340 - 335}{1.7} = 2.941$ (3 d.p.)

Using the tables in Appendix 2, this gives

$$P(X + Y > 340) = 1 - 0.9984 = 0.0016$$

∴ The required probability is 0.0016.

Example

The weights of male and female students at a college may be assumed to be normally distributed, the males' weights having mean 72 kg and standard deviation 6 kg and the females' weights having mean 60 kg and standard deviation 8 kg. If one male and one female student are chosen at random calculate the probabilities that:

(a) the female is heavier than the male

(b) the female's weight is at least $\dfrac{3}{4}$ of the male's weight.

Solution

In symbol form we have

$M \sim N(72, 6^2)$ for the males and $F \sim N(60, 8^2)$ for the females

(a) It follows that $\qquad M - F \sim N(72{-}60, 6^2 + 8^2)$

(note the signs)

i.e. $M - F \sim N(12, 100)$

We therefore expect the male to be 12 kg heavier with a standard deviation of $\sqrt{100} = 10$ kg. If we require the female to be heavier, then we need to find the shaded area in Fig. 1.2:

Figure 1.2

The required z value is $\dfrac{0 - 12}{10} = -1.2$ and this gives

$$P(M - F < 0) = 1 - 0.8849 = 0.1151$$

∴ There is a 11.51% chance of the female being heavier.

(b)   This time we need the distribution of $F - \frac{3}{4}M$

$$\therefore \quad F - \frac{3}{4}M \sim N\left(60 - \frac{3}{4} \times 72,\ 8^2 + \left(\frac{3}{4}\right)^2 \times 6^2\right)$$

(Note, carefully, the variance.)

$$\therefore \quad F - \frac{3}{4}M \sim N(6,\ 84.25)$$

Since the question asks for $P(F - \frac{3}{4}M > 0)$, we need to find the shaded area in Fig. 1.3:

Figure 1.3

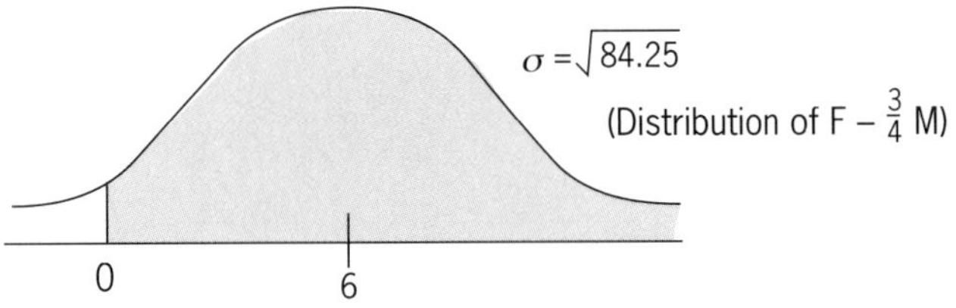

$\sigma = \sqrt{84.25}$

(Distribution of $F - \frac{3}{4}M$)

The required z value is $\dfrac{0 - 6}{\sqrt{84.25}} = -0.653$ (3 d.p.)

and this gives $P(F - \frac{3}{4}M > 0) = 0.7432$

$\therefore$ the required probability is 0.7432.

Example

Milk chocolates have weight which are normally distributed with mean 22 g and standard deviation 0.3 g. Plain chocolates have weights which are normally distributed with mean 17 g and standard deviation 0.2 g. An empty chocolate box has a normally distributed weight with mean 35 g and standard deviation 0.15 g.

(a)   Agatha buys a box of chocolate which contains six milk chocolates and four plain chocolates. What is the chance that Agatha's box of chocolates weighs more than 236 g?

(b)   What is the chance that the weights of any two milk chocolates differ by at least 1 g?

Solution

In symbol form we have

$M \sim N(22,\ 0.3^2)$ for milk chocolate,

$P \sim N(17,\ 0.2^2)$ for plain chocolate and

$B \sim N(35,\ 0.15^2)$ for the empty box.

(a)   The box of chocolates is given by the sum of eleven random variables, i.e.

$$B + M_1 + M_2 + M_3 + M_4 + M_5 + M_6 + P_1 + P_2 + P_3 + P_4.$$

The box of chocolates will *not* be $B + 6M + 4P$ because we are not, for example, taking one milk chocolate and multiplying its weight by 6. We are taking six separate milk chocolates.

Now $E(B + M_1 + M_2 + M_3 + M_4 + M_5 + M_6 + P_1 + P_2 + P_3 + P_4)$

$$= 35 + 6 \times 22 + 4 \times 17 = 235$$

and $Var(B + M_1 + M_2 + M_3 + M_4 + M_5 + M_6 + P_1 + P_2 + P_3 + P_4)$

$$= 0.15^2 + 0.3^2 + 0.3^2 + 0.3^2 + 0.3^2 + 0.3^2 + 0.3^2 + 0.2^2 + 0.2^2 + 0.2^2 + 0.2^2$$

$$= 0.7225$$

∴ The weight of Agatha's box of chocolates is given by N(235, 0.7225) and we now have to find the shaded area below:

Figure 1.4

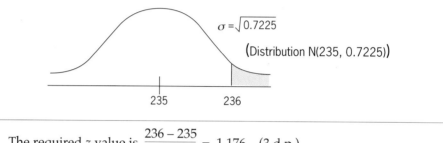

$\sigma = \sqrt{0.7225}$

(Distribution N(235, 0.7225))

235   236

The required z value is $\dfrac{236 - 235}{\sqrt{0.7225}} = 1.176$ (3 d.p.)

and this gives the required probability as $1 - 0.88 = 0.12$

∴ There is a 12% chance that Agatha's box of chocolates weighs more than 236 g.

(b) For the difference between two milk chocolates we have to consider the random variable $M_1 - M_2$.

Now $E(M_1 - M_2) = 22 - 22 = 0$ and $Var(M_1 - M_2) = 0.3^2 + 0.3^2 = 0.18$

∴ We now have to find the shaded area below:

Figure 1.5

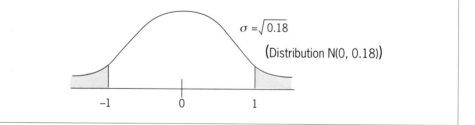

$\sigma = \sqrt{0.18}$

(Distribution N(0, 0.18))

−1   0   1

[Note that we have to find the area to the left of −1 as well as the area to the right of 1.]

A required z value is $\dfrac{1 - 0}{\sqrt{0.18}} = 2.357$ (3 d.p.)

which gives the required probability as $2 \times (1 - 0.9908) = 0.0184$

∴ There is a 1.84% chance that the weight of two milk chocolates differ by at least 1 g.

Practice questions B

1 The weights (g) of three types of slugs are defined by the independent random variables A, B and C where

   $A \sim N(12, 4)$, $B \sim N(14, 9)$ and $C \sim N(16, 16)$

   The following questions all refer to the above distributions.

   (a) Two A-type slugs are chosen at random. What is the chance that their total weight exceeds 26 g?

   (b) Two A-type slugs are chosen at random. What is the chance that their weights differ by no more than 3 g?

   (c) An A-type slug and a B-type slug are chosen at random. What is the chance that the B-type slug is heavier then the A-type slug?

   (d) An A-type slug and a B-type slug are chosen at random. What is the chance that their total weight is no more than 29 g?

   (e) Two B-type slugs are chosen at random. What is the chance that their weights differ by at least 2 g?

   (f) An A-type slug, a B-type slug and a C-type slug are chosen at random. What is the chance that their total weight is less than 30 g?

   (g) An A-type slug and a C-type slug are chosen at random. What is the chance that the A-type slug is the heavier?

   (h) A B-type slug and a C-type slug are randomly chosen. What is the chance that the C-type slug is the lighter?

   (i) An A-type slug and a B-type slug are randomly chosen. What is the chance that the B-type slug is at least 5 g heavier than the A-type slug?

   (j) Three B-type slugs and four A-type slugs are randomly chosen. What is the chance that the total weight of the four A-type slugs is greater than the total weight of the three B-type slugs?

   (k) An A-type slug, a B-type slug and two C-type slugs are randomly chosen. What is the chance that the A-type and B-type slugs together weigh more than the two C-type slugs?

2 The arrival time of the Ipswich train is normally distributed about a mean of 10.08 with a standard deviation of 3 minutes. The arrival time of the Sheringham train is independently normally distributed about a mean of 10.20 with a standard deviation of 4 minutes.

   Find the chance that:

   (a) the Ipswich train arrives before the Sheringham train

   (b) the Ipswich train arrives at least 8 minutes before the Sheringham train

   (c) the difference in arrival times between the Ipswich and Sheringham trains is no more than 4 minutes.

3 The weight of luggage that aircraft passengers take with them is normally distributed with mean 20 kg and standard deviation 5 kg. A certain type of aircraft carries 100 passengers. What is the probability that the total weight of the passengers' luggage exceeds 2150 kg?

SUMMARY EXERCISE

1 X and Y are independent random variables such that:

   $E(X) = 20$, $Var(X) = 2$, $E(Y) = 24$, $Var(Y) = 3$.

   Evaluate the following:

   (a) $E(5X + 7)$

   (b) $Var(5X + 7)$

   (c) $E(5X + 7Y)$

   (d) $Var(5X + 7Y)$

   (e) $E(5X - 7Y)$

   (f) $Var(5X - 7Y)$

   (g) $E(5 - 7Y)$

   (h) $Var(5 - 7Y)$

2 A random variable R takes the integer value r with probability p(r) where

   $$p(r) = kr^3 \quad r = 1, 2, 3, 4$$
   $$p(r) = 0 \quad \text{otherwise}$$

   Find:

   (a) the value of k and display the distribution on graph paper

   (b) the mean and variance of the distribution

   (c) the mean and variance of $5R - 3$.

3   Discrete random variables X and Y have distributions given in the tables below:

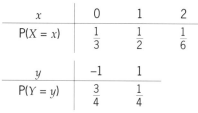

x	0	1	2
$P(X = x)$	$\frac{1}{3}$	$\frac{1}{2}$	$\frac{1}{6}$

y	−1	1
$P(Y = y)$	$\frac{3}{4}$	$\frac{1}{4}$

Find the distributions of:

(a) $3X$

(b) $2Y$

(c) $3X + 2Y$.

and find:

(d) $E(X)$, $Var(X)$

(e) $E(Y)$, $Var(Y)$

(f) $E(3X + 2Y)$, $Var(3X + 2Y)$

Verify the results that:

$E(3X + 2Y) = 3E(X) + 2E(Y)$

$Var(3X + 2Y) = 9Var(X) + 4Var(Y)$

4   The mass, at harvest, of the fruit of a particular tree has been found to have a normal distribution. Records show that of 200 fruits harvested, 20 were below 125 g in mass and 40 above 155 g.

(a) What is your estimate of the mean and standard deviation of the distribution?

Fruits from another tree have masses that are normally distributed with mean 133 g and standard deviation 10 g. Four fruits chosen at random from the second tree are made up into packets.

(b) What percentage of packets will be less than 500 g in mass?

5   The diameters of axles supplied by a factory have a mean value of 19.92 mm and a standard deviation of 0.05 mm. The inside diameters of bearings supplied by another factory have a mean of 20.04 mm and a standard deviation of 0.03 mm. What is the mean and standard deviation of the random variable defined to be the diameter of a bearing less than the diameter of an axle?

Assuming that both dimensions are normally distributed, what percentage of axles and bearings taken at random will not fit?

6   A doctor working in a clinic finds that the consulting times of his patients are independently normally distributed with mean 5 minutes and standard deviation 1.5 minutes. He sees his patients consecutively with no gaps between them, starting at 10 am.

(a) At what time should the tenth patient arrange to meet a taxi after their consultation, so as to be 99% certain that he will not keep it waiting?

(b) If the doctor sees 22 patients in all, what is the probability that he will finish before noon?

SUMMARY

In this section we have seen that, for random variables X and Y:

- $E(aX + bY) = aE(X) + bE(Y)$
- $E(aX - bY) = aE(X) - bE(Y)$
- $E(a) = a$ (for constants a and b)

Also, if **X and Y are independent** then:

- $Var(aX + bY) = a^2 Var(X) + b^2 Var(Y)$
- $Var(aX - bY) = a^2 Var(X) + b^2 Var(Y)$ [Note the sign.]
- $Var(a) = 0$

For **independent normal variables** X_1 and X_2, where $X_1 \sim N(\mu_1, \sigma_1^2)$ and $X_2 \sim N(\mu_2, \sigma_2^2)$:

- $X_1 + X_2 \sim N(\mu_1 + \mu_2, \sigma_1^2 + \sigma_2^2)$
- $X_1 - X_2 \sim N(\mu_1 - \mu_2, \sigma_1^2 + \sigma_2^2)$ [Note the sign.]
- $aX_1 + bX_2 \sim N(a\mu_1 + b\mu_2, a^2\sigma_1^2 + b^2\sigma_2^2)$
- $aX_1 - bX_2 \sim N(a\mu_1 - b\mu_2, a^2\sigma_1^2 + b^2\sigma_2^2)$ [Note the sign.]

In particular, if $A \sim N(\mu, \sigma^2)$, then:

● *two members of A will be*

$A_1 + A_2 \sim N(2\mu, 2\sigma^2)$

but

● *a number of A doubled will be*

$2A \sim N(2\mu, 2^2\sigma^2)$ [Note the 2^2.]

i.e. $2A \sim N(2\mu, 4\sigma^2)$.

ANSWERS

Practice questions A

1 (a) 50, 9   (b) 10, 9
      (c) 80, 21   (d) −10, 24
      (e) 50, 4   (f) 50, 56
      (g) 90, 9   (h) 170, 61

2 (a) 37, 5   (b) −62, 59

3 (a) $3\frac{1}{2}$, $2\frac{11}{12}$ (b) $10\frac{1}{2}$, $8\frac{3}{4}$
      (c) 13, $17\frac{1}{2}$

4 (a) 70, $58\frac{1}{3}$ (b) 70, $1166\frac{2}{3}$

5 (a) $(X_1 + X_2)$ ∴ 96, 10
      (b) $(2X)$ ∴ 96, 20

6 (a) Var $X = 16$ ∴ 80, 400
      (b) 172, 988

7 (a) 112, 16
      (b) 88, 16

8 (a) 3, 0.75, 0.0375
      (b) (i) 1.5, 0.075   (ii) 5, 0.075

Practice questions B

1 (a) $A_1 + A_2$ ∴

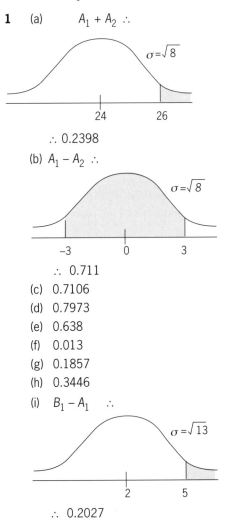

$\sigma = \sqrt{8}$

24   26

∴ 0.2398

(b) $A_1 - A_2$ ∴

$\sigma = \sqrt{8}$

−3   0   3

∴ 0.711

(c) 0.7106
(d) 0.7973
(e) 0.638
(f) 0.013
(g) 0.1857
(h) 0.3446
(i) $B_1 - A_1$ ∴

$\sigma = \sqrt{13}$

2   5

∴ 0.2027

(j) $A_1 + A_2 + A_3 + A_4 - B_1 - B_2 - B_3$

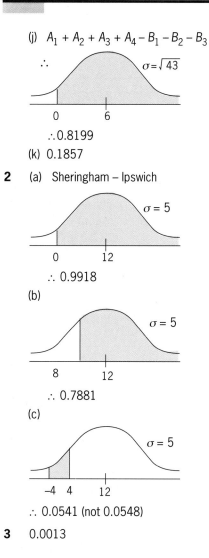

∴ $\sigma = \sqrt{43}$

     0   6

∴0.8199

(k) 0.1857

2   (a)   Sheringham – Ipswich

$\sigma = 5$

     0   12

∴ 0.9918

(b)

$\sigma = 5$

     8   12

∴ 0.7881

(c)

$\sigma = 5$

   −4   4   12

∴ 0.0541 (not 0.0548)

3   0.0013

Sampling

INTRODUCTION In previous sections we have assumed 'random samples' but haven't really indicated how such samples might be achieved. We aim to put that right in this section. Furthermore, we'll look at other types of sampling – stratified, systematic and quota – with a view to deciding which is the best to use in particular circumstances.

For this part of the 'A' level course you are expected to write a few lines of English (part of the so-called 'Key Skills' requirements) and to use a bit of common sense. You may say 'but that's not maths!', but if the sampling is faulty, then so are all the subsequent mathematical deductions. For this reason samples must be very carefully chosen.

Random sampling

OCR S2 5.12.4 (a),(b)

This is by far the most important method of taking samples and simply requires that each member of the population has an equal chance of appearing in the sample. This type of sample can most easily be obtained if there exists a list of all members of the population (**a sampling frame**) and items from the list can be selected in a random unsystematic way (or **without bias**, in statistician's language).

Very often in the sampling process a decision has to be made about whether to allow an item to be chosen repeatedly. If we replace an item after sampling so it has a chance of being selected again, we refer to this as simple random sampling **with replacement**. Sampling **without replacement** happens when an item is not replaced, i.e. once it has been selected, it has no further chance of being selected. Which method we choose depends on the nature of the problem under investigation. For large populations, the distinction between the two methods is minimal, but if the population is small, then sampling without replacement could affect the random nature of the sample.

As a simple illustration of how a random sample might be obtained, consider the problem of obtaining a sample of ten dates from the days of a year (not a leap year). The dates from 1st January to 31st December could be written on to cards, placed in a container, shuffled and then 10 cards could be selected from the container. No particular date is more likely to turn up than any other and each has a chance of $\frac{1}{365}$ of being selected.

A useful technique for larger populations is to use tables of random numbers (see Appendix 1 at the end of the book). These tables consist of blocks of digits, 0 to 9, which have been generated in such a way that there is no bias towards any of the ten digits. The blocks have no particular significance – they are simply there to help with reading the tables. Most random number tables are generated by a computer program and some calculators can generate three-digit random numbers by a similar means.

The normal format for a random number generated by a calculator is as a three-digit decimal. By ignoring the decimal point (to which no significance should be attached), these numbers can just be used as random sequences consisting of three digits. To generate a random number consisting of five digits for example, it would be perfectly valid to generate two random numbers via the calculator and ignore the third digit of the second number.

Returning to the problem of selecting 10 dates from the year, how might random numbers help here?

We need to ensure that each date has an equal chance of occurring. It would therefore be *incorrect* to number the days as follows:

1	1st January
2	2nd January
.	
.	
.	
364	30th December
365	31st December

the reason being that whereas the digit 1 has a probability of $\frac{1}{10}$ of appearing, the sequence of digits 365 has a probability of $\frac{1}{1000}$ $(=\frac{1}{10} \times \frac{1}{10} \times \frac{1}{10})$ of appearing so we would produce a sample very biased towards dates early in the year. So we must ensure as a first step that each random number we associate with a given date has the same chance of appearing. The obvious way in this example is to attach a three-digit number to each date (if the tables are truly random then each three-digit sequence is as likely as any other).

A suitable numbering system would therefore be:

001	1st January
002	2nd January
.	
.	
.	
364	30th December
365	31st December

We now proceed systematically through the random number table and see what turns up. If the sequence 000 or any sequence greater than 365 turns up, we simply reject it and move to the next three-digit sequence. Starting (randomly) at the 3rd line of the tables (we can start anywhere we like), gives the following sequence:

784, 803, 378, 226, 010, 664, 659, etc.

You will notice that we are going to reject a lot of number triples. In fact only 2 out of the first 7 are acceptable for our scheme – 226 and 010.

We can increase the economy of this method in several ways and one way is as follows:

1st January	001	401
2nd January	002	402
	.	.
	.	.
	.	.
30th December	364	764
31st December	365	765

This simple step halves the number of rejected triples and does not affect the random nature of the sample obtained.

Out of the seven triples above we can now make use of 4, giving

784 – reject
803 – reject
378 – reject
226 – 14th August
010 – 10th January
664 – 21st September
659 – 16th September

and we can continue in this way until we have a sample of the required size.

It is quite likely in an example of this type that we would want a sample of 10 different dates, so we would be finding a simple random sample without replacement. In effect, this would mean that if a date happened to come up a second time, we would simply reject it and move to the next random number triple.

Practice questions A

1 Choose three random letters of the alphabet by starting in the 1st line of the tables.

2 Choose 4 random years from the 20th century by starting in the 2nd line of the tables.

3 Choose 5 random years from the 19th and 20th centuries by starting in the 3rd line of the tables (You will need to continue into the 4th line.)

4 The population of a town is 5123. By starting in the 4th row of the tables, choose a random sample of 5. (You'll need to use line 5 as well.)

5 By starting in the 5th line, choose 4 different random numbers from the population 0, 1, 2, 3, 4, 5, 6, 7, 8 and 9.

6 Choose 3 random people from a population of 55 million. Start on line 6.

7 A class consists of 20 girls and 14 boys. Choose a random sample of 3 girls and 2 boys by numbering the class as 1, 2, … , 20 for the girls and then 21, 22, …, 34 for the boys. Start on line 6.

8 Choose a random sample of 5 from the population −3, 0, 4, 4, 11, 11 and 28. Start on line 8.

9 Is the following a random sample?

(a) From a bag containing 30 identical looking marbles, John picks out two marbles, one by one, without replacement. **C** 3.2

(b) From the same bag, Jenny picks out three marbles with replacement.

(c) From the days of the week – Monday and Wednesday.

10 How do you select a random sample of 5 in the following cases:

(a) Out of 450 pupils 5 are to be selected for a committee. **C** 3.2

(b) Customers coming out of a store in a steady stream?

Stratified sampling

It may happen that a population under examination falls naturally into sub-groups or *strata*. As a simple example, the population of students in a college is made up of male and female students or full-time and part-time students. Consider a college with 500 registered students who group according to the table:

	Full-time	Part-time
Male	150	100
Female	120	130

If we take a random sample of 50 from the group then we could by chance pick out only full-time students or only male students (each of the $^{300}C_{50}$ possible selections is equally likely under random sampling). We may conduct the sampling when there are proportionately fewer part-time students on campus – quite likely as, by definition, part-time students are at the college less of the time. This would introduce a bias which may be undesirable. **Stratified sampling** provides a means of ensuring that the sampling is in strict proportion to the numbers in each of the 'strata'.

In this case we would divide up our sample of 50 in the ratio
$150 : 100 : 120 : 130$

giving   15   full-time males
              10   part-time males
              12   full-time females
              13   part-time females

and the sample would more properly reflect the natural sub-divisions occurring within the population.

We could use simple random sampling to select the required numbers of students in the sub-categories.

The major advantage of stratified sampling is that the final result – the sample – should reflect any natural sub-division within the population. There are problems associated with this method of sampling, however:

- Firstly, there has to be sufficient information about the population to enable the person taking the sample to delineate the sub-divisions. Such information may not be readily accessible or it may be confidential.

- It is most important that the sub-divisions do not overlap and that they include the whole population. This may involve considerable organisational work, particularly for large populations.

Systematic sampling

In this type of sampling, the candidates for the sample are chosen according to some regular pattern, such as:

- pick out every 10th member from a list of a population

- call at every house numbered 13 or 27 from the houses on an estate

- select every 300th item from a factory production line for quality control testing.

One advantage of such a method is that it is a fairly simple procedure to carry out – for example, a computer program could probably control the third example given above. This method also has the obvious advantage of being simple to explain to somebody who might be employed in taking a sample and therefore may reduce the costs of taking a sample.

A major disadvantage is that the samples obtained are no longer truly random and therefore the data obtained from them must be used with care. It may be, for example, in the factory production line that every 300th item produced is faulty as a result of some fault in the machinery which produces a fault in a regular way, e.g. if every nth item is faulty and n is a factor of 300, then this would not be picked up by the sampling procedure.

Practice questions B   C 3.2

1 A school of 600 pupils has 400 boys and 200 girls. In order to seek opinions on school food, a sample of 12 pupils is to be selected.

(a) Give a reason why a stratified sample might be preferable to a random sample.

(b) If you took a stratified sample, how many boys would it contain and how many girls would it contain?

(c) Choose the girls by using line 1 of the random number table.

(d) Choose the boys by using line 3 of the random number table.

2 The following table gives the age distribution of pupils in an 'A' level maths sixth-form group.

Age (yr)	Boys	Girls	Totals
15	8	17	25
16	9	16	25
17	9	15	24
Totals	26	48	74

You are asked to choose a representative sample of 9 pupils.

(a) Give a reason for choosing a stratified sample in preference to a random sample.

(b) Explain how you would decide how many boys and girls you should choose from each age group.

(c) Choose your sample, using random number tables.

3 Customers are coming out of a store in a steady stream and you are asked to take a sample.

(a) Why might a systematic sample be preferable?

(b) How would you choose such a sample?

4 An hotel has 8 floors with 15 rooms on each floor. You are asked to seek out the residents' opinions by means of a sample. Describe how you would choose:

(a) a random sample

(b) a stratified sample

(c) a systematic sample.

Which sampling procedure might be preferable? Give your reasons.

Quota sampling

In this type of sampling the population is stratified as before but *the choice of which person to interview is left to the interviewer.* Quota sampling and stratified sampling therefore differ in the way in which each strata sample is chosen – in stratified sampling, it is chosen using random number tables but in quota sampling, it's up to the interviewer to decide. Quota sampling is therefore much quicker to carry out but bias might well be introduced by the interviewer's preferences!

Practice questions C 🔑 C 3.2

1 A class of 30 pupils contains 20 boys and 10 girls. You wish to choose a representative committee of 3 pupils. Describe how you would choose

(a) random sample

(b) a stratified sample

(c) a quota sample.

Which sampling procedure might be preferable?

2 A fruiterer is offered consignments of apples, pears, oranges, peaches, melons and pineapples. Before accepting delivery he decides to sample the fruits for suitability. What sampling procedure is preferable?

3 'Out of 350 cars parked in a car park, 48 were fitted with an anti-theft device on the steering wheels. Assuming that the cars form a random sample of parked cars ...'

Give a reason why the assumption of randomness in the above question might not be valid.

4 In order to carry out a statistical test on the leaves of a tree, a student examines 150 leaves that are within reach.

Comment on the validity of the sampling method chosen.

5 'A thesis claimed 42% of people ... In order to test the claim, a sceptical reader asked 14 of her friends and ...'

Give a reason why the sceptical reader's conclusion might be invalid.

6 In order to assess Mary's skills at dart throwing, John analysed her first 20 throws. Comment on John's approach.

7 The manager of a supermarket wishes to judge the effect of a new layout on the customers. On the day that the new layout was introduced she asked the first 100 customers whether or not they approved of the new layout.

Comment on the way in which the sample was chosen, and suggest a method of obtaining a more representative sample.

8 There are 8 packets of cereal on a supermarket's shelf. Give a description of a suitable population from which the 8 packets could be a sample.

9 Give one advantage and one disadvantage with taking a sample.

10 Comment on the reliability of the following ways of finding a sample

(a) Decide whether the potatoes have cooked properly by testing one with a fork.

(b) Choosing a staff representative by picking a name out of a hat.

(c) Find out about the most popular make of car by counting 20 cars in a town's car park.

11 How would you choose a sample from a lorry load of potatoes?

12 It is required to choose a random person from a village. It is decided to do this by choosing a house at random and then choosing a member of that household at random.

Explain why this method will not produce a random person from the village.

SUMMARY EXERCISE

1 (a) State a situation in which you would consider using:

(i) a systematic sample

(ii) a stratified sample

when sampling from a population.

Give a specific example in each case.

(b) Give one advantage and one disadvantage associated with stratified sampling.

2 A college of 3000 students has students registered in four departments, Arts, Science, Education and Crafts. The Principal wishes to take a sample from the student population to gain information about likely student response to a rearrangement of the college timetable so as to hold lectures on Wednesday, previously reserved for sports.

What sampling method would you advise the Principal to use? Give reasons to justify your choice.

3 A manufacturer of a new cereal wishes to predict the likely volume of sales in a town. Four schemes, as below, are proposed for selecting a suitable sample of people to question. Which is best and why?

(a) Choose people at random from the phone book.

(b) Select houses at random and choose a random person from each house chosen.

(c) Take every 40th person whose name appears on the electoral register.

(d) Stand on a street corner and choose a quota sample according to sex, age, etc.

SUMMARY

In this section we have seen:

- how to use random number tables when taking a **random sample**

- that in a random sample every member of the population has an equal chance of being chosen

- that a random sample might miss out chunks of the population

- that a **stratified sample** attempts to ensure a good representation of the population

- that a **quota sample** is quicker to assemble than a stratified sample but it can possibly lead to interviewer bias

- that a **systematic sample** could be ideal when the population size isn't exactly known but, once again, it might lead to bias if not carefully used, e.g. all odd numbers will only give houses on one side of the street.

ANSWERS

Practice questions A

1 Let $A = 01$, $B = 02$, ... , $Z = 26$

The table gives 13, 10 and 07, i.e. M, J and G.

2 Let $1900 = 00$, $1901 = 01$, ... , $1999 = 99$

The table gives 1960, 1978, 1948 and 1912.

3 Let $1800 = 000$, $1801 = 001$, $1802 = 002$, ... $1900 = 100$, $1901 = 101$, ... , $1999 = 199$.

The table gives 010, 172, 193, 079 and 041, i.e. 1810, 1972, 1993, 1879 and 1841.

4 1840, 4151, 2504, 2071 and 3925.

5 9, 0, 3 and 2 (Only count 9 once.)

6 53589636, 11784023 and 31775394

7 $G_1 = 01$, $G_2 = 02$, ... , $G_{20} = 20$,

$B_1 = 21$, ... , $B_{14} = 34$

$\therefore$ $32 = B_2$, $12 = G_{12}$, $17 = G_{17}$, $16 = G_{16}$, $23 = B_3$. (Girls 2, 15 and 11 are ignored.)

8 Let $-3 = 1$, $0 = 2$, $4 = 3$, $4 = 4$, $11 = 5$, $11 = 6$ and $28 = 7$

Table gives *3, 1, 7, 5* and *4.*

$\therefore$ Sample is 4, −3, 28, 11 and 4.

9 (a) Yes – provided John doesn't cheat.

(b) Yes.

(c) No, unless Monday and Wednesday have been randomly obtained.

10 (a) Call them 001, 002, ... , 450 and use a random number table, choosing blocks of three digits.

(b) Estimate the likely total number of customers and take that estimate as your population. Then choose the sample using random number tables. (An alternative method involves systematic sampling – read on!)

Practice questions B

1 (a) A random sample might produce all boys and their eating preferences may be different from girls!

   (b) 8 boys and 4 girls

   (c) Call the girls in the school 001, 002, ... 200
      Sampling gives 100, 192, 060 and 133

   (d) Call the boys in the school 001, 002, ..., 400
      Sampling gives 378, 226, 010, 172, 193, 249, 079 and 041.

2 (a) A random sample might produce all 15-year-olds.

   (b)

Age	Boys	Girls	
15	1	2	(The ratio of boys to girls
16	1	2	is approximately 1 : 2 in
17	1	2	each age group)

   (c) Choose a sample of 1 from 8, 1 from 9 and 1 from 9 and then for the girls choose 2 from 17, 2 from 16 and 2 from 15.

3 (a) Random and stratified sampling would be awkward because the exact number of customers cannot be known in advance.

   (b) Depending on likely numbers, you might choose (say) every 50th customer. That should give you time to get your responses before moving on to the next person.

4 (a) Number the residents as 001, 002, ... and then use random number tables to choose the sample.

   (b) Choose a random number of residents from each floor.

   (c) Sample every 10th (say) resident as they left the hotel.

Stratified sampling probably preferable. A random sample might be too clustered and a systematic sample might not give a good cross-section of residents.

Practice questions C

1 (a) Use random number tables to choose a sample of 3 from 1, 2, 3, ... , 20, 21, ... 30.

   (b) Choose 2 random boys from 20 and then 1 random girl from 10.

   (c) Choose 2 boys and 1 girl – the choice is yours.

Stratified sampling probably best. Random sampling might choose all boys and quota sampling would be more likely to include friends than enemies!

2 A quota sample would do very well – test one or two fruits of each variety. A stratified sample would be impractical – it would take too long. A random sample might produce all melons!

3 Probably only the expensive parked cars will have an anti-theft device.

4 The population of leaves *beyond* the student's reach has been excluded. The test is probably invalid.

5 She only sampled her friends.

6 Mary will only be 'warming up'. A biased sample of results.

7 The first 100 customers will not be representative of all the supermarket's customers – mid-day and late-night shoppers will be unrepresented for a start.

   A systematic sample might be more appropriate – choose every 100th customer for example.

8 The 8 packets might be a sample of that week's delivery.

9 Advantage – it is quick.

   Disadvantage – you might omit an important sub-set of the population.

10 (a) OK providing the potatoes are all more or less the same size

   (b) OK providing the staff are all 'similar'

   (c) A town's cars would be unrepresentative of both a city's and a village's cars. No good.

11 Assuming the potatoes are stored in crates, use a systematic sampling method to select the crates (every 5th, say) and then pick one potato out at random (a quota sample) from each crate.

12 A person in a household of one will stand more chance of being chosen than a person in a household of four. Hence the method does not produce a random person.

SECTION 3

Estimation, confidence intervals and tests

INTRODUCTION

If we took as our population all the males in the UK, and then a random sample of these males, how could we work out the following:

(a) a best estimate for the mean weight of males in the UK

(b) a best estimate of the standard deviation of this population of males

(c) an interval in which we could be 95% sure that the population mean weight of the males will be situated

(d) the chance that the mean weight of the male population in the UK exceeds 90 kg?

If we then took a sample of females, how could we test the hypothesis that the mean weight of the female population is less than the mean weight of the male population or, indeed, is at least 10 kg less?

It is questions such as these that this section is aiming to answer.

Sampling distributions of statistics

OCR S2 5.12.4 (c)

If we take a sample from a population with (unknown) mean μ, then a natural choice for an *estimate* of the value of μ, the population mean, is the calculated value of $\bar{x}$, the sample mean. $\bar{x}$ is called a **point estimate** for μ.

If we were to take many samples of a given size from the population, we would be surprised if each sample gave the same value for $\bar{x}$. Indeed if we were to do this, we could obtain a whole range of values for $\bar{x}$, draw up a frequency table of the values and calculate an mean value of all of the $\bar{x}$'s. We would be treating $\bar{x}$ as if it were itself a value from a new random variable $\bar{X}$.

The random variable $\bar{X}$ is called an **estimator** of the population mean μ, and its specific value $\bar{x}$ calculated from the sample is called an **estimate** of the population mean μ.

The distribution of the random variable $\bar{X}$ is called the **sampling distribution of the mean** (or alternatively the **distribution of the sample mean**), and it has some useful and remarkable properties.

Throughout the following we will consider a population having mean μ and variance σ^2 (not necessarily known to us) and a sample of n independent observations (a random sample of size n) where for each observation X_i

$$E(X_i) = \mu \quad \text{and} \quad \text{Var}(X_i) = \sigma^2$$

i.e. each observation is a value of a random variable with the same distribution as the population from which it is drawn.

Let $\bar{X} = \dfrac{1}{n}(X_1 + X_2 + \ldots + X_n)$

21

then the mean of this distribution is

$$E(\overline{X}) = E\left[\frac{1}{n}(X_1 + X_2 + \dots + X_n)\right]$$

$$= \frac{1}{n}E(X_1 + X_2 + \dots + X_n)$$

$$= \frac{1}{n}\left(E(X_1) + E(X_2) + \dots + E(X_n)\right)$$

giving $E(\overline{X}) = \dfrac{1}{n}(\mu + \mu + \dots + \mu)$

$$= \frac{1}{n}(n\mu) = \mu$$

In other words the expected or mean value of the sample mean is precisely equal to the population mean.

$\overline{X}$ is, as a result of this property, called an **unbiased estimator** of μ and correspondingly $\overline{x}$, the value of $\overline{X}$ which we find for a particular sample, is called an **unbiased estimate** of μ.

We can also think about the variance of the random variable $\overline{X}$.

$$Var(\overline{X}) = Var\left[\frac{1}{n}(X_1 + X_2 + \dots + X_n)\right]$$

$$= \frac{1}{n^2}\left[Var(X_1) + Var(X_2) + \dots Var(X_n)\right]$$

$$= \frac{1}{n^2}(\sigma^2 + \sigma^2 + \dots + \sigma^2)$$

$$= \frac{n\sigma^2}{n^2} = \frac{\sigma^2}{n}$$

i.e. the variance of $\overline{X}$ is equal to the variance of the population divided by the sample size.

The square root of this, the standard deviation of $\overline{X}$, is called the **standard error** of the mean.

So far we have:

$\overline{X}$ is the sample mean

$E(\overline{X}) = \mu$, the population mean

$Var(\overline{X}) = \dfrac{\sigma^2}{n}$, the population variance divided by the sample size

$$\text{Standard error of } \overline{X} = \sqrt{Var(\overline{X})} = \frac{\sigma}{\sqrt{n}}$$

The standard error is used extensively.

Practice questions A

1 X is a random variable and a sample of size n produces a random variable $\overline{X}$. In each of the following cases find (or write down) the values of

(i) $E(X) = M$ (ii) standard deviation of X

(iii) $E(\overline{X})$ (iv) standard error of $\overline{X}$. $\sqrt{\dfrac{\sigma^2}{n}}$

(a) $X \sim N(50, 9)$ and $n = 4$
(b) $X \sim N(100, 25)$ and $n = 36$
(c) $X \sim B(50, 0.2)$ and $n = 25$
(d) $X \sim B(100, 0.4)$ and $n = 9$
(e) $X \sim P(6)$ and $n = 4$
(f) $X \sim P(5)$ *rectangular* and $n = 2$
(g) $X \sim U(3, 5)$ and $n = 4$
(h) $X \sim U(-1, 1)$ and $n = 25$

2 X is a random variable which is normally distributed with mean 20 and variance 4.

A sample of size 9 is taken and the mean $\overline{X}$ of this sample is calculated. Find $E(\overline{X})$ and $Var(\overline{X})$.

3 X is normally distributed with mean 40 and variance 9. A sample of size n is taken and the mean of the sample is calculated. What is the expected value of this mean? If the mean is $\overline{X}$ and $Var(\overline{X}) = 0.18$, what is n?

4   If $X \sim N(30, 32)$, write down the expectation and variance of the random variable $\overline{X}$, where $\overline{X}$ is the mean of 10 independent observations of X.

The Central Limit theorem

OCR S2 5.12.4 (d),(e)

We have seen in section 1 that a sum of *independent* normal random variables is itself a normal variable.

$\therefore$ If $X_1 \sim N(\mu_1, \sigma_1{}^2)$, $X_2 \sim N(\mu_2, \sigma_2{}^2) \ldots X_n \sim N(\mu_n, \sigma_n{}^2)$

then $X_1 + X_2 + \ldots + X_n \quad \sim N(\mu_1 + \mu_2 + \ldots + \mu_n, \sigma_1{}^2 + \sigma_2{}^2 + \ldots + \sigma_n{}^2)$

$$\sum_{i=1}^{n} X_i \sim N\left(\sum_{i=1}^{n} \mu_i, \sum_{i=1}^{n} \sigma_i^2\right)$$

Now if our parent population is a normal distribution and we think of our random sample as a series of independent normal random variables with

$E(\overline{X}) = \mu$ and $Var(\overline{X}) = \dfrac{\sigma^2}{n}$

then the distribution of $\overline{X}$ will itself be a normal distribution according to the result above (each observation being of a normal variable).

Hence if the sample is drawn from a population with a normal distribution we have

$$\overline{X} \sim N\left(\mu, \frac{\sigma^2}{n}\right)$$

and this result is true for any sample size n.

A more powerful result (the derivation of which is not required for 'A' level) states that if $Y = X_1 + X_2 + X_3 + \ldots + X_n$ and n is 'sufficiently large' then Y is approximately normal regardless of the distributions of the individual X_i's. This is a very powerful result known as the **Central Limit theorem** and for our purposes it enables us to conclude that:

$$\bar{X} \sim N\left(\mu, \frac{\sigma^2}{n}\right)$$

- this result is *true* for all $n \geq 1$ if samples are drawn from a *normal* population

- this result is *approximately true* for 'sufficiently large' n if the samples are drawn from a population which is *not normal.*

The phrase 'sufficiently large' is not an easy one to put actual values to as it will depend on the nature of the distribution. For a binomial distribution with $p = \frac{1}{2}$ the value of n could be fairly small for the approximation to be good since with $p = \frac{1}{2}$ the binomial distribution has the same symmetry as the normal distribution.

As p gets further away from $\frac{1}{2}$ so n would have to be correspondingly larger for a good approximation. In general the more skewed a distribution, the larger n would have to be for an approximation to be a good one. It may be assumed in examination questions that the values of the sample size given are sufficiently large for the result to apply.

| **Example** | If $X \sim N(15, 25)$ and a sample of size 10 is drawn from this distribution, find the probability that sample mean is between 14 and 16.5. |

| **Solution** | Since the question is asking about the sample mean we need to consider the distribution |

$$\bar{X} \sim N\left(15, \frac{25}{10}\right)$$

and for this distribution we find

$$P(14 < \bar{X} < 16.5)$$

By the usual standardisation procedure we have

$$P(14 < \bar{X} < 16.5) = P\left(\frac{14 - 15}{\sqrt{\frac{25}{10}}} < z < \frac{16.5 - 15}{\sqrt{\frac{25}{10}}}\right)$$

$$= P(-0.6325 < Z < 0.9487)$$

$$= 0.565 \ (3 \ \text{d.p.})$$

| **Example** | If $X \sim B(300, 0.7)$ and a sample of size 50 is taken from this distribution, estimate the probability that the sample mean is between 205 and 213. |

| **Solution** | $n = 50$ is sufficiently large for the Central Limit theorem to apply. |

The mean of $X = 300 \times 0.7 = 210$

and the variance of $X = 300 \times 0.7 \times 0.3 = 63$

$$\therefore \quad \bar{X} \sim N\left(210, \frac{63}{50}\right)$$

and so $\quad P(205 < \overline{X} < 213) \;=\; P\left(\dfrac{200-205}{\sqrt{\dfrac{63}{50}}} < z < \dfrac{213-210}{\sqrt{\dfrac{63}{50}}}\right)$

$$= P(-4.4544 < z < 2.6726)$$

$$= 0.9962$$

Practice questions B

1 A random sample of size 15 is taken from a normal distribution with mean 60 and standard deviation 4. Find the probability that the mean of the sample is less than 58.

2 If $X \sim N(21, 90)$ and a sample of size 10 is taken, find the probability that $\overline{X}$ is between 18 and 27.

3 If $X \sim N(200, 80)$ and a random sample of size 5 is taken from the distribution, find the probability that the sample mean is greater than 207.

4 A random sample of size 100 is taken from $B(20, 0.6)$. What is the chance that the sample mean exceeds 12.4?

5 A random sample of size 30 is taken from $P(4)$. Find the chance that the sample mean is less than 4.5.

6 Describe the distribution of $\overline{X}$ in the following cases:
   (a) $X \sim N(60, 25)$ and $n = 225$
   (b) $X \sim P(3)$ and $n = 400$
   (c) $X \sim B(10, 0.6)$ and $n = 160$

An unbiased estimator of population variance

OCR S2 5.12.4 (f)

We noted on p. 22 that $E(\overline{X}) = \mu$ and we referred to this property by saying that $\overline{X}$ is an unbiased estimator of μ.

Now it turns out that the quantity $\dfrac{\sum(x_i - \overline{x})^2}{n}$ does *not* have the property that its expectation equals σ^2.

On average it under-estimates the value of σ^2 and is therefore a biased estimator of σ^2. A minor adjustment provides us with an estimator s^2 which does have the desired property of being unbiased, i.e. of giving good results *on average*.

$$s^2 = \left(\frac{n}{n-1}\right)\frac{\sum(x_i - \overline{x})^2}{n}$$

is an unbiased estimator for σ^2, where n is the size of the sample.

So, to provide us with an unbiased estimator of σ^2, first find $\dfrac{\sum(x_i - \overline{x})^2}{n}$ or equivalently and more simply, $\dfrac{\sum x_i^2}{n} - \overline{x}^2$, and then multiply by the factor $\dfrac{n}{n-1}$.

It should be noted that if n is large then multiplying by this factor will not make much difference,

e.g. if $n = 500$, then we would work out $\dfrac{500}{499} \times \dfrac{\sum(x_i - \overline{x})^2}{500}$.

However in examinations it is correct to use this factor anyway.

Practice questions C

1 For the following sets of sample data, find unbiased estimates for the population mean and variance

   (a) 19.3, 19.61, 18.27, 18.9, 19.14, 19.90, 18.76, 19.10

   (b) $n = 34$, $\sum x = 330$, $\sum x^2 = 23700$

   (c)

x	$0 \le x < 4$	$4 \le x < 8$	$8 \le x < 12$	$12 \le x < 16$	$16 \le x < 20$
f	3	6	24	10	7

2 A sample of rats is caught and weighed as follows:

Weight (kg)	1.4	1.6	1.8	2.0	2.2
Frequency	3	8	16	11	2

Find unbiased estimates for the mean weight of a rat and its standard error.

3 What is meant by 'unbiased estimate'?   **C** 3.2

Confidence intervals for the population mean   OCR S3 5.13.3 (a)

We'll now use the Central Limit Theorem to calculate intervals which contain μ with a certain probability.

As before we have $X \sim N(\mu, \sigma^2)$ with means of samples of size n producing $\bar{X} \sim N\left(\mu, \dfrac{\sigma^2}{n}\right)$. Therefore all sample means $\bar{x}$ will belong to the distribution shown in Fig. 3.1.

Figure 3.1

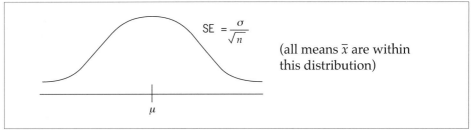

(all means $\bar{x}$ are within this distribution)

We can be 95% sure (for example) that all sample means $\bar{x}$ will be within the limits A and B shown in Fig. 3.2.

Figure 3.2

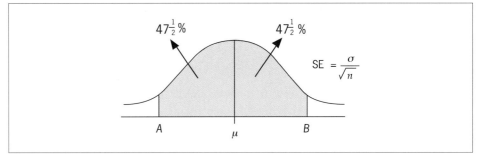

But the limit B is given by

$$\frac{B - \mu}{\frac{\sigma}{\sqrt{n}}} = 1.96 \;\Rightarrow\; B = \mu + \frac{1.96\sigma}{\sqrt{n}}.$$

Similarly the limit A is given by

$$\frac{A - \mu}{\frac{\sigma}{\sqrt{n}}} = -1.96 \implies A = \mu - \frac{1.96\sigma}{\sqrt{n}}$$

∴ We can be 95% sure that our sample mean $\bar{x}$ will satisfy

$$\mu - \frac{1.96\sigma}{\sqrt{n}} \le \bar{x} \le \mu + \frac{1.96\sigma}{\sqrt{n}}$$

$$\implies \quad -\frac{1.96\sigma}{\sqrt{n}} \le -\mu + \bar{x} \le \frac{1.96\sigma}{\sqrt{n}}$$

$$\implies \quad -\bar{x} - \frac{1.96\sigma}{\sqrt{n}} \le -\mu \le -\bar{x} + \frac{1.96\sigma}{\sqrt{n}}$$

$$\implies \quad \bar{x} + \frac{1.96\sigma}{\sqrt{n}} \ge \mu \ge \bar{x} - \frac{1.96\sigma}{\sqrt{n}}$$

The interval $\bar{x} \pm \dfrac{1.96\sigma}{\sqrt{n}}$ is called the *95% confidence interval* for the population mean μ.

And in summary form:

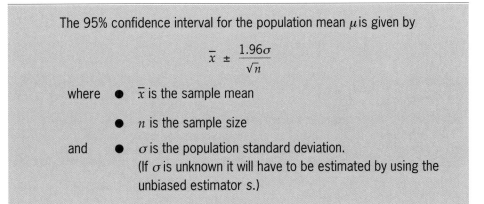

The 95% confidence interval for the population mean μ is given by

$$\bar{x} \pm \frac{1.96\sigma}{\sqrt{n}}$$

where
- $\bar{x}$ is the sample mean
- n is the sample size

and
- σ is the population standard deviation.
  (If σ is unknown it will have to be estimated by using the unbiased estimator s.)

Before, we had a point estimate of μ, namely $\bar{x}$. Now we have an interval estimate of μ.

| **Example** | A sample of size 15 from a normal population with unknown mean μ and known variance 25, gives a value of $\bar{x} = 32$. Give a 95% confidence interval for μ. |

| **Solution** | Quoting the result above (which is all you will be expected to do in the exam) gives |

$$32 \pm 1.96\frac{5}{\sqrt{15}} = 32 \pm 2.53$$

$$= [29.47, 34.53]$$

i.e. $P(29.47 \le \mu \le 34.53) = 0.95$

It is important to realise what this result means. Suppose we took a different sample, then it is highly likely that a different value of $\bar{x}$ would be obtained and therefore a different confidence interval. Confidence intervals are, therefore, *variable quantities* dependent on which sample we happen to select. However, μ is a fixed quantity for a population.

The information which a confidence interval provides is therefore that there is a 95% probability that the interval contains μ.

Example	A sample of size 2000 is taken from a population with unknown distribution, unknown mean μ and unknown variance σ^2.

Calculation gives $\sum x_i = 21\,310$, $\sum x_i^2 = 458\,650$.

Find a 95% confidence interval for μ.

Solution	$n = 2000$ is sufficiently large to assume that

$$\bar{X} \sim N\left(\mu, \frac{\sigma^2}{n}\right) \text{ applies}$$

In this example the population variance is unknown and therefore we have to use the information from the sample to find an estimate for it.

The sample mean $= \bar{x} = \dfrac{21\,310}{2000} = 10.66$ (2 d.p.)

and the sample variance $= \dfrac{\sum x_i^2}{n} - \bar{x}^2 = 115.70$ (2 d.p.)

and therefore $s^2 = \dfrac{2000}{1999} \times 115.70 = 115.75$ (2 d.p.)

The confidence interval is therefore

$$10.66 \pm 1.96 \times \frac{\sqrt{115.75}}{\sqrt{2000}} = 10.66 \pm 0.47$$

$$= [10.19, 11.13]$$

i.e. $P(10.19 \le \mu \le 11.13) = 0.95$

For certain purposes it may be necessary to have an interval within which we would like to be 99% confident or perhaps 90% confident that it contains μ. Usually 95% intervals are used, but in principle we can find a confidence interval of any width.

Example	For the data in the previous example calculate a 99% confidence interval for μ.

Solution	The interval is going to be of the form

$$\bar{x} \pm z \frac{s}{\sqrt{n}}$$

i.e. the only change is in the multiple of $\dfrac{s}{\sqrt{n}}$, the standard error of $\bar{x}$.

To find z we need to know the solution to $P(-z < Z < z) = 0.99$ and from the normal distribution tables (looking up an area of 99.5%), we find $z = 2.575$.

Our interval is therefore:

$$10.66 \pm 2.575 \times \frac{10.76}{\sqrt{2000}} = 10.66 \pm 0.62 \text{ (2 d.p.)}$$

$$= [10.04, 11.28] \text{ (2 d.p.)}$$

i.e. $P(10.04 \leq \mu \leq 11.28) = 0.99$

Note that the 99% interval is wider than the 95% interval for μ. To increase our level of confidence we need to widen the interval.

Practice questions D

1   The 95% confidence limits for the population mean μ are given by

$$\bar{x} \pm \frac{1.96\sigma}{\sqrt{n}}.$$

Write down similar expressions for:

(a) the 99% confidence limits

(b) the 90% confidence limits

(c) the 98% confidence limits,

of the population mean μ.

2   A sample of worms had lengths as follows (in cm):

9.5, 9.5, 11.2, 10.6, 9.9, 11.1, 10.9, 9.8, 10.1, 10.2, 10.9 and 11.0

Assuming that the sample came from a normal distribution with standard deviation 2, find a 95% confidence interval for the mean length of all worms.

3   A random sample of 120 measurements gives:

$n = 120, \ \sum x = 1008, \ \sum x^2 = 8640$

Find:

(a) a 97% confidence interval

(b) a 99% confidence interval

for the population mean.

4   A normal distribution has a variance of 9. A sample from this distribution was: 11, 10, 15, 4 and 16. Calculate a best estimate for the population mean and give a 95% confidence interval for this best estimate.

5   A sample of 16 values of X is taken from $N(\mu, 3)$ and is found to have a mean $\bar{X} = 4.2$, Write down the distribution of $\bar{X}$. Hence find a 95% confidence interval for the population mean.

6   A random sample of 100 values of X was obtained with the following results:

$\sum x = 960, \quad \sum x^2 = 14\,625.$

Estimate the mean and variance of the population from which this sample was taken.

Find a 95% confidence interval for the mean of X.

Determination of sample size   OCR S3 5.13.3 (a)

One useful variant on the confidence interval calculation relates to the determination of the sample size.

| **Example** | An electronics firm manufactures a printed circuit board which is to be installed in another component. We wish to estimate μ (the mean length of the circuit board) to within limits of 0.05 mm. If we know that the standard deviation of the length is 0.5 mm, determine the minimum sample size required if we require a probability of 95% |

Solution

We have:

$$\bar{x} \pm 1.96 \frac{\sigma}{\sqrt{n}}$$

and we require $\bar{x} \pm 0.05$ mm. This implies that:

$$1.96 \frac{\sigma}{\sqrt{n}} = 0.05 \text{ mm}$$

Given that $\sigma = 0.5$ mm, it follows that

$$\frac{1.96(0.5)}{\sqrt{n}} = 0.05 \Rightarrow \frac{1.96(0.5)}{0.05} = \sqrt{n} \Rightarrow 19.6 = \sqrt{n} \Rightarrow n = 384.16$$

which must be rounded to 385 to give the required interval size of within 0.05 mm.

Practice questions E

1 A sample of size n is taken from a normal distribution with mean 50 and variance 4. If the probability that the mean of the sample exceeds 51 is 0.025, find the value of n.

2 $X \sim N(50, 9)$ and a sample of size n produces a mean $\bar{X}$. If the chance of $\bar{X}$ exceeding 49.5 is 0.88, find the value of n.

3 A large number of random samples of size n are taken from the distribution of X where $X \sim N(74, 36)$ and the sample means are calculated.

   If $P(\bar{X} > 72) = 0.854$, estimate the value of n.

4 A normal distribution has a mean of 30 and a variance of 5. Find n such that the probability that the mean of n observations exceeds 30.5 is less than 0.01.

5 The 95% confidence limits for the mean of a certain population are 1230 to 1270. This interval is based on the results of a sample of 25. Deduce the best estimates for:

   (a) the population mean

   (b) the population standard deviation.

6 A sample of 100 gave the 95% confidence limits for the population mean as 150 ± 1.8. What are the 90% confidence limits for the population mean?

7 Based on a sample of 500, the 95% confidence interval for the population mean had a width of 20. What is the population standard deviation?

Use of Central Limit theorem in hypothesis testing   OCR S2 5.12.4 (i)

We noted earlier that if X is any random variable (discrete or continuous) and our sample size n is sufficiently large, then the variable $\bar{X}$ will have an approximately normal distribution with parameters μ and $\frac{\sigma^2}{n}$ where μ and σ^2 are the mean and variance of X.

i.e. $\bar{X} \approx N\left(\mu, \frac{\sigma^2}{n}\right)$.

This powerful theorem enables us to conduct hypothesis tests about mean values even if we don't know the nature of the original distribution.

Indeed, it may be that σ^2 is also unknown. In these circumstances it would be necessary to approximate the value of σ^2 by s^2, the unbiased estimator of σ^2.

More generally then, we would have

$$\bar{X} \approx N\left(\mu, \frac{s^2}{n}\right)$$

We would then use the sample of values to find $\bar{x}$ and s^2 and proceed to test the value of $\bar{x}$ against the value of μ using the test statistic

$$z = \frac{\bar{x} - \mu}{\dfrac{s}{\sqrt{n}}}$$

Let's look at a couple of examples.

Example

The lengths of metal bars produced by a machine are normally distributed with mean length 420 cm and standard deviation 12 cm. The machine is serviced, after which a sample of 100 bars gives a mean length of 423 cm. Is there evidence, at the 5% level, of a change in the mean length of the bars produced by the machine, assuming that the standard deviation remains the same?

Solution

We begin with: $H_0 : \mu = 420$

$$AH : \mu \neq 420$$

(The key word in the question is 'changed'.)

For a sample of size 100 we would expect $\bar{X}$ to be given by

$$E(\bar{X}) = 420 \text{ and } Var(\bar{X}) = \frac{144}{100} = 1.44$$

∴ The means of samples of size 100 should have the following distribution:

Figure 3.3

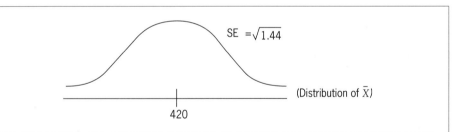

The significance level of the test is 5% and so we want to know whether our sample reading of 423 is in the shaded area below:

Figure 3.4

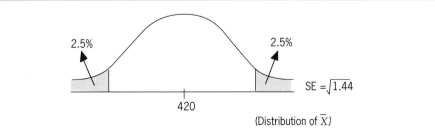

∴ We need to find $P(\bar{X} \geq 423)$ and the required z value is $\dfrac{423 - 420}{\sqrt{1.44}} = 2.5$.

The normal tables then give us $P(\bar{X} \geq 423) = 1 - 0.9938 = 0.0062$

i.e. there is only a 0.62% chance that the sample mean exceeds 423.

∴ 423 is in the critical region.

∴ We accept AH.

∴ The mean length of bar has changed.

Example

A normal distribution is thought to have a mean of 50. A random sample of 100 gave a mean of 52.6 and a standard deviation of 14.5. Is there evidence at the 5% level that the mean has increased?

Solution

We begin with: $H_0 : \mu = 50$

$$AH : \mu > 50$$

(The key word in the question is 'increased')

The sample standard deviation is 14.5 and so our best estimate for the population standard deviation is

$$14.5 \sqrt{\frac{100}{99}} = 14.57 \ (2 \ d.p.)$$

∴ For a sample of size 100 we would expect $\bar{X}$ to be given by $E(\bar{X}) = 50$
and $Var(\bar{X}) = \frac{14.57^2}{100} = 2.12 \ (2 \ d.p.)$

∴ The means of samples of size 100 should have the following distribution:

Figure 3.5

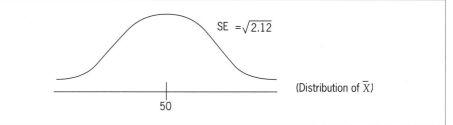

The significance level of the test is 5% and so we want to know whether our sample reading of 52.6 is in the shaded area below:

Figure 3.6

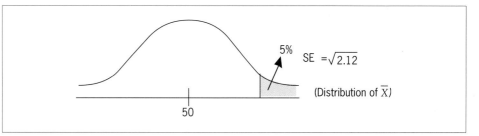

∴ We need to find $P(\bar{X} \geq 52.6)$ and the required z value is
$$\frac{52.6 - 50}{\sqrt{2.12}} = 1.784.$$

The normal tables then give us $P(\bar{X} \geq 52.6) = 1 - 0.9628 = 0.0372$

i.e. there is only a 3.72% chance that the sample mean exceeds 52.6

$\therefore$ 52.6 is in the critical region.

$\therefore$ We accept AH.

$\therefore$ The mean has increased.

Practice questions F   🔑 C 3.2

1 Experience has shown that the scores obtained in a particular test are normally distributed with mean score 70 and variance 36. When the test is taken by a random sample of 36 students, the mean score is 68.5. Is there sufficient evidence, at the 3% level, that these students have not performed as well as expected?

2 A company manufacturers rope whose breaking strengths have a mean of 300 lb and a standard deviation of 24 lb. A sample of 64 ropes was tested and their mean breaking strength was 306 lbs. Test at the 1% level of significance whether the mean breaking strength has increased.

3 Testing $H_0 : \mu = 20$ against AH: $\mu \neq 20$, the test statistic $Z = 3.1$ was calculated. What is your conclusion at the 1% level?

4 The national scores in IQ tests are normally distributed with mean 100 and standard deviation 15. A local school tested a random sample of 10 teachers and got the following results: 119, 131, 95, 107, 125, 90, 123, 89, 103 and 103. Is there evidence at the 5% level of significance that the teachers are any different from the national population?

5 A sample of 40 observations from a normal distribution gave $\sum x = 24$ and $\sum x^2 = 596$. Test, at the 5% level, whether the mean of the distribution is zero.

6 The mean number of matches in a box is 50. A sample of 150 boxes purchased in a town gave

$$\sum x = 7464 \quad \text{and} \quad \sum x^2 = 371793$$

Using a 5% level of significance, is the mean number of matches per box in the town significantly different from expected?

7 The lifetime (in days) of a sample of 100 light bulbs were as follows:

Lifetime (days)	4	8	12	16	20
Frequency	10	17	47	20	6

Test the hypothesis that the population mean is 12 days. (Use a 5% level of significance.)

8 A random variable X is distributed as $N(\mu, 4)$. A sample of 36 readings gave

$$\sum x = 462 \quad \text{and} \quad \sum x^2 = 6074$$

Find: (a)   a best estimate for μ

         (b)   the 99% confidence limits for μ.

Also test the hypothesis $\mu = 12.2$ against the alternative hypothesis $\mu > 12.2$, using a 5% level of significance.

9 The mean life of dogs is normally distributed and the local vet claims that the mean lifetime is 9.5 years. I think it is different.

Among 16 dogs I found lifetimes of

11, 12, 12, 12, 11, 10, 8, 4, 10, 11, 11, 12, 9, 9, 11 and 14

Test the vet's claim at the 2.5% level, clearly stating H_0, AH and your conclusion.

10 I carry out an hypothesis test at the 1% level with $H_0 : \mu = 80$ and AH : $\mu < 80$. If the test statistic $Z = 2.3$, what is my conclusion?

11 What is meant by a significance test at the 5% level?

12 Mice have a mean weight of 150 g; their standard deviation is unknown. With the change in farming techniques, it is thought that the mean weight of mice has changed. A sample of 50 mice gave me

$$\sum x = 7473 \quad \text{and} \quad \sum x^2 = 1116956.5$$

(a) Find a best estimate for the population standard deviation.

(b) Carry out an hypothesis test at the 4% level of significance, stating clearly H_0, AH and your conclusion.

Testing the difference between two sample means

OCR S3 5.13.4 (b)

It is also possible to apply the same principles to a test involving two samples/populations. Consider the following.

Example

A dietician is testing two different diets on volunteers with a view to assessing weight loss under each diet regime. Under Diet *A*, 120 people took part and achieved an average weight loss of 450 grams, standard deviation 87 grams. For Diet *B*, 150 people took part and achieved an average weight loss of 425 grams, standard deviation 95 grams.

Is there any difference, at the 5% significance level, in the effect of these two diets on weight loss?

Solution

First we find best estimates for the population standard deviation. They are:

$$S_1 = 87 \ \sqrt{\frac{120}{119}} \ = 87.36 \ \text{(2 d.p.) for Diet A}$$

and $\quad S_2 = 95 \ \sqrt{\dfrac{150}{149}} \ = 95.32$ (d.p.) for Diet B

Since both samples are large, the Central Limit Theorem tells us that

$$\bar{A} \sim \text{N}\left(\mu_a, \frac{87.36^2}{120}\right) \qquad \text{i.e.} \quad \bar{A} \sim \text{N}(\mu_a, 63.605)$$

and $\quad \bar{B} \sim \text{N}\left(\mu_b, \dfrac{95.32^2}{150}\right) \qquad$ i.e. $\quad \bar{B} \sim \text{N}(\mu_b, 60.570)$

It follows that $\ \bar{A} - \bar{B} \sim \text{N}(\mu_a - \mu_b, 124.175)$

Now take $\qquad \text{H}_0 : \mu_a = \mu_b$

and $\qquad\qquad \text{AH} : \mu_a \neq \mu_b$

(The key word in the question is 'difference'.)

$\therefore \quad$ If H_0 is true, $\bar{A} - \bar{B} \sim \text{N}(0, 124.175)$

and we will reject H_0 if the sample difference of $450 - 425 = 25$ is in the shaded area shown in Fig. 3.7:

Figure 3.7

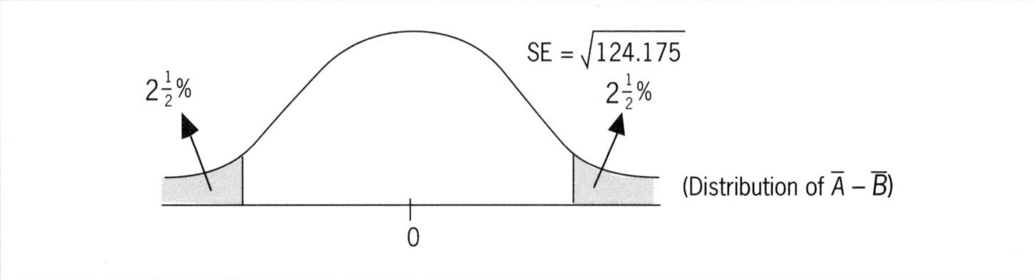

(Distribution of $\bar{A} - \bar{B}$)

The test statistic $z = \dfrac{25 - 0}{\sqrt{124.175}} = 2.243$, which gives us

$\text{P}(\bar{A} - \bar{B} \geq 25) = 1 - 0.9876 = 0.0124$, i.e. there is only a 1.24% chance of getting a mean difference of 25 or more.

$\therefore \quad$ 25 is in the critical region.

$\therefore \quad$ We accept AH.

$\therefore \quad$ There is a difference between the mean weight loss under the two diets.

Practice questions G 🔧 C 3.2

1 A test at the 5% level of significance is to be carried out for $H_0 : \mu_1 = \mu_2$ against AH : $\mu_1 \neq \mu_2$, where μ_1 and μ_2 are the means of independent normal distributions whose standard deviations are the means of independent normal distributions when standard deviations are 4.986 and 3.524. A random sample of 25 observations from the first distribution had a mean of 80.9, while an independent random sample of 36 observations from the second distribution had a mean of 78.4. What is the conclusion of the test?

2 Two types of battery were compared for the length of time they lasted. The data obtained are summarised in the table below.

Battery type	Sample size	Sample mean	Sample SD
A	200	1990	25.5
B	150	2000	32.8

Test the hypothesis that the populations from which the samples were drawn have equal means against the alternative hypothesis of unequal means. Use a 1% level of significance.

3 The data obtained from the weights of children are summarised in the table below.

	Sample size	Sample mean
Boys	64	9.4
Girls	81	9.0

It is known that the weights are normally distributed and that the population standard deviations for boys and girls are 0.894 and 1.095 respectively. Test at the 5% level the hypothesis that there is no difference between the average weights of boys and girls.

4 The daily outputs of factories A and B are independent normal distributions with means 12 and 15 tonnes and standard deviations 0.5 and 0.6 tonnes respectively.

It is felt that staff changes may affect the means (whilst leaving standard deviations unchanged). A check over 100 days gave means of 12 and 15.18 tonnes respectively.

Test at the 5% level whether or not the difference in the means is different from the value which would have been expected.

5 It is found that over a certain period at one telephone exchange 200 subscribers taken at random made a total of 13,248 calls. During the same time, a random sample of 300 subscribers at another exchange made a total of 20,922 calls. The standard deviation of the number of calls made by a subscriber at either exchange in the period is 8. Is there any evidence of a difference between the subscribers at the two exchanges in their average frequency of calls?

6 Samples of leaves were collected from two oak trees A and B. The number of galls was counted on each leaf and the mean and standard deviation of the number of galls per leaf was calculated with results as follows:

Tree	Sample size	Sample mean	Sample SD
A	60	11.4	2.6
B	80	10.7	3.1

Assuming normal distributions, do the data provide evidence at the 5% level of different population means for the two trees?

7 The lengths (in mm) of nine screws selected at random from a large consignment are found to be:

7.99, 8.01, 8.00, 8.02, 8.03, 7.99, 8.00, 8.01 and 8.01

Calculate unbiased estimates of the population mean and variance.

From a second large consignment, sixteen screws are selected at random and their mean length (in mm) is found to be 7.992. Assuming a normal distribution with variance 0.0001, test, at the 5% level, the hypothesis that this population has the same mean as the first population against the alternative hypothesis that this population has a smaller mean than the first population. (You may assume that the first population is also normal.)

8 Heights (mm) are normally distributed and sample data for males and females is as follows:

Sex	Sample size	$\sum x$	$\sum x^2$
Male	120	198	327
Female	160	248	385

Find best estimates for the population mean and variance of

(a) males and (b) females

Test the hypothesis (at the 1% level) that the mean height of the population of males exceeds the mean height of the population of females by 0.08 m.

SUMMARY EXERCISE

1 If $n = 2000$, $\bar{x} = 10.66$ and $s = 10.76$, find a 90% confidence interval for μ.

2 A machine is regulated to dispense liquid into cartons in such a way that the amount of liquid dispensed on each occasion is normally distributed with a standard deviation of 20 ml.

Find 99% confidence limits for the mean amount of liquid dispensed if a random sample of 40 cartons has an average content of 266 ml.

3 A sample of ten packets of sugar was chosen at random and the contents of each packet were weighed giving the following results in grams:

998.4, 1002.3, 999.2, 1001.5, 997.6, 999.4, 1002.8, 1001.5, 999.5, 1002.6.

Assuming that the sample comes from a population having a normal distribution find:

(a) A 95% confidence interval for μ

(b) A 99% confidence interval for μ

(c) The number of packets which would need to be sampled to give a 95% confidence interval of width less than 1.2.

4 The random variable X is normally distributed with mean μ and variance σ^2.

(a) Write down the distribution of the sample mean $\bar{X}$ of a random sample of size n.

An efficiency expert wishes to determine the mean time taken to drill a fixed number of holes in a metal sheet.

(b) Determine how large a random sample is needed so that the expert can be 95% certain that the sample mean time will differ from the true mean time by less than 15 seconds. Assume that it is known from previous studies that $\sigma = 40$ seconds.

5 A food processor produces large batches of jars of jam. In each batch the gross weight of a jar is known to be normally distributed with standard deviation 7.5 g.

The gross weights, in grams, of a random sample from a particular batch were:

517, 481, 504, 482, 503, 497, 512, 487, 497, 503, 509

(a) Calculate a 90% confidence interval for the mean gross weight of this batch.

(b) The manufacturer claims that the mean gross weight of a jar in a batch is at least 502 g. Test this claim at the 5% significance level.

(c) Explain why, if the manufacturer had claimed that the mean gross weight was at least 496 g, no further calculations would be necessary to test this claim.

6 A population X has $E(X) = 50$ and $Var(X) = 4$.

A sample of size n is taken and the mean $\bar{X}$ is calculated. If the chance that $\bar{X}$ exceeds 50.2 is 0.0228, what is n?

7 Consider the sample data below:

Sector	Sample size	Sample mean	Sample SD
Public	200	95	19
Private	300	101	16

Test at the 5% level whether public sector wages are lower than private sector wages.

In this section we have seen that:

- for any random variable X, the **sampling distribution** of the means of samples of size n is denoted by $\bar{X}$

- if $E(X) = \mu$ and $\text{Var}(X) = \sigma^2$

  then

  $E(\bar{X}) = \mu$ and $\text{Var}(\bar{X}) = \dfrac{\sigma^2}{n}$.

- $\dfrac{\sigma}{\sqrt{n}}$ = the standard deviation of $\bar{X}$ = the **standard error of the mean**

- if $X \sim N(\mu, \sigma^2)$ then $\bar{X} \sim N\left(\mu, \dfrac{\sigma^2}{n}\right)$

  but

  if X is *not* normal then $\bar{X} \approx N\left(\mu, \dfrac{\sigma^2}{n}\right)$, provided that n is large (the **Central Limit Theorem**).

- the **best estimate for the population mean** μ is $\bar{x}$ (the sample mean)

- the **best estimate for the population variance** σ^2 is $\left(\dfrac{n}{n-1}\right)s^2$,

  where s^2 is the sample variance

- **95% confidence limits** for the population mean μ are $\bar{x} \pm \dfrac{1.96\sigma}{\sqrt{n}}$

- **99% confidence limits** for the population mean μ are $\bar{x} \pm \dfrac{2.575\sigma}{\sqrt{n}}$

- the distribution of $\bar{X}$ is used when **testing hypotheses involving means**, i.e. $\bar{X} \sim N\left(\mu, \dfrac{\sigma^2}{n}\right)$

- **hypothesis testing** involving the difference between two sample means uses the following:

  $$\bar{X}_1 \sim N\left(\mu_1, \dfrac{\sigma_1^2}{n_1}\right) \text{ and } \bar{X}_2 \sim N\left(\mu_2, \dfrac{\sigma_2^2}{n_2}\right)$$

  $$\Rightarrow \bar{X}_1 - \bar{X}_2 \sim N\left(\mu_1 - \mu_2, \dfrac{\sigma_1^2}{n_1} + \dfrac{\sigma_2^2}{n_2}\right)$$

ANSWERS

Practice questions A

1 (a) $E(X) = 50$, SD = 3, $E(\bar{X}) = 50$, SE = 1.5

(b) $E(X) = 100$, SD = 5, $E(\bar{X}) = 100$, SE = $\frac{5}{6}$

(c) $E(X) = 10$, SD = $\sqrt{8}$, $E(\bar{X}) = 10$,

SE $= \frac{\sqrt{8}}{5} = 0.4\sqrt{2}$

(d) $E(X) = 40$, SD = $\sqrt{24}$, $E(\bar{X}) = 40$,

SE $= \frac{\sqrt{24}}{3} = \frac{2\sqrt{2}}{\sqrt{3}}$

(e) $E(X) = 6$, SD = $\sqrt{6}$, $E(\bar{X}) = 6$,

SE $= \frac{\sqrt{6}}{2} = \sqrt{1.5}$

(f) $E(X) = 5$, SD = $\sqrt{5}$, $E(\bar{X}) = 5$,

SE $= \frac{\sqrt{5}}{\sqrt{2}} = \sqrt{2.5}$

(g) $E(X) = 4$, SD = $\sqrt{\frac{1}{3}}$, $E(\bar{X}) = 4$,

SE $= \frac{1}{2\sqrt{3}} = \frac{\sqrt{3}}{6}$

(h) $E(X) = 0$, SD = $\sqrt{\frac{1}{3}}$, $E(\bar{X}) = 0$,

SE $= \frac{1}{5\sqrt{3}} = \frac{\sqrt{3}}{15}$

2 20, $\frac{4}{9}$

3 40, 50

4 30, 3.2

Practice questions B

1

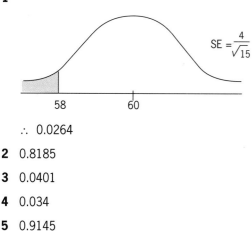

SE $= \frac{4}{\sqrt{15}}$

∴ 0.0264

2 0.8185

3 0.0401

4 0.034

5 0.9145

6 (a) $\bar{X}$ is normally distributed with mean 60 and variance $\frac{1}{9}$

(b) $\bar{X}$ is *approximately* normally distributed with mean 3 and variance 0.0075

(c) $\bar{X}$ is *approximately* normally distributed with mean 6 and variance 0.015.

Practice questions C

1 (a) 19.1225, 0.25459

(b) 9.71, 621.12 (2 d.p.)

(c) 10.96, 17.35 (2 d.p.)

2 1.805 kg, 0.1999 kg

3 An 'unbiased estimate' of a population parameter is the most likely value of that parameter.

Practice questions D

1 (a) $\bar{x} \pm 2.575 \frac{\sigma}{\sqrt{n}}$ (Look up 0.995)

(b) $\bar{x} \pm 1.645 \frac{\sigma}{\sqrt{n}}$ (Look up 0.95)

(c) $\bar{x} \pm 2.326 \frac{\sigma}{\sqrt{n}}$ (Look up 0.99)

2 9.26 to 11.52

3 (a) 8.161 to 8.638 (b) 8.118 to 8.682

4 11.2, 13.83 to 8.57

5 $\bar{X} \sim N\left(4.2, \frac{3}{16}\right)$, 3.35 to 5.05

6 9.6, 54.64; 8.15 to 11.05

Practice questions E

1 15.4 ∴ 16

2 49.7 ∴ 50

3 10

4 $n > 108$

5 (a) 1250 (b) 51.0

6 150 ± 1.51

7 114.09

Practice questions F

1 $P(\bar{X} \le 68.5) = 0.0668$
Accept H_0. No evidence of poor performance.

2 $P(\bar{X} \ge 306) = 0.0228$
Accept H_0. No change in mean breaking strength.

3 Since 0.1% < 0.5%, reject H_0.
Accept that $\mu \ne 20$.

4 $P(\bar{X} \ge 108.5) = 0.0367$ and 3.67% > 2.5%
Accept H_0. Teachers no different

5 (Best estimate of $\sigma = 3.8617$.)
$P(\bar{X} \ge 0.6) = 0.164$.
Accept H_0 that mean is zero.

6 $P(\bar{X} \le 49.76) = 0.0336$ and 3.36% > $2\frac{1}{2}$%.
Accept H_0 ∴ no change in mean.

7 $P(\bar{X} \le 11.8) = 0.309$.
Accept H_0 ∴ population mean = 12

8 (a) $12\frac{5}{6}$ (b) 13.69 to 11.975
2.87% < 5% ∴ accept AH that $\mu > 12.2$

9 (Best estimate for $\sigma = 2.25$)
$H_0 : \mu = 9.5$ $AH : \mu \ne 9.5$
$P(\bar{X} \ge 10.4375) = 0.0478$ and 4.78% > 1.25%.
Accept H_0. Vet correct.

10 1.07% > 1% ∴ just accept H_0

11 There is a 5% chance that you reject H_0 when it is true.

12 (a) 0.9249379
(b) $H_0 : \mu = 150$, $AH : \mu \ne 150$.
$P(\bar{X} \le 149.46) = 0$ (off the table).
Accept AH. Mean weight of mice has changed.

Practice questions G

1 $P(\bar{A} - \bar{B} \ge 2.5) = 0.0154$ and 1.54% < 2.5%.
Accept AH ∴ $\mu_1 \ne \mu_2$

2 (SE = $\sqrt{10.488}$)
$P(\bar{A} - \bar{B} \ge 10) = 0.001$ and 0.1% < 0.5%
Accept AH ∴ $\mu_1 \ne \mu_2$

3 (SE = $\sqrt{0.0273}$)
Since 0.78% < 2.5%
Accept AH ∴ $\mu_1 \ne \mu_2$

4 (SE = $\sqrt{0.0061}$)
Since 1.07% < 2.5%.
Accept AH so difference in means is no longer 3 tonnes.

5 (SE = $\sqrt{0.533}$)
Since $P(\bar{A} - \bar{B} \ge 3.5)$ gives a z value of 4.79
Accept AH so that exchanges are different.

6 (SE = $\sqrt{0.2362}$)
Since 7.49% > 2.5%
Accept H_0. No difference between the means.

7 8.006 and 0.000175
(SE = $\sqrt{0.0000257}$)
Since 0.19% < 5%
2nd population has smaller mean than the first.

8 (a) 1.65, 0.00252 (b) 1.55, 0.00377
(SE = $\sqrt{0.0000446}$)
Since 0.13% < 0.5%
Accept AH ∴ means don't differ by 0.08

SECTION 4

Goodness of fit and contingency tables

INTRODUCTION In this section we'll be considering the feasibility of fitting particular statistical models to a given set of randomly collected data. The models that we shall be looking at are discrete uniform, binomial and Poisson distributions, as well as the continuous normal and uniform distributions.

In general we are going to be given a set of sample values ('observed' values) and then, having selected a possible model, compare them with a set of 'expected' values. The procedure that enables us to compare these two sets of values is known as the chi-square test (written χ^2 test).

The χ^2 distribution

OCR S3 5.13.5 (a),(b)

We state that the distribution has a PDF of the form:

$$f(x) = A_v \left(\frac{x}{2}\right)^{\frac{v}{2}-1} e^{-\frac{x}{2}} \text{ for } x > 0$$

where A is a constant dependent on v and v is an integral parameter known as the number of degrees of freedom of the distribution. (v is another Greek letter, nu, pronounced 'new'.) You do not need to understand the meaning of A, and v will be explained as we use it.

Fig. 4.1 illustrates the nature of the PDF for certain values of v.

Figure 4.1

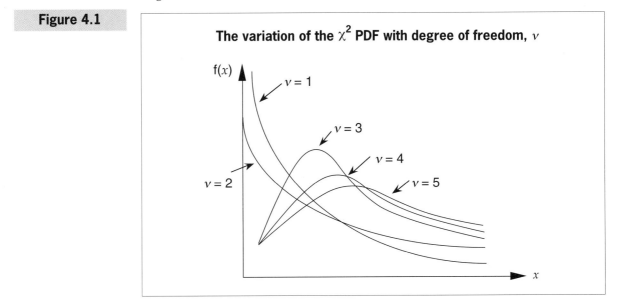

The variation of the χ^2 PDF with degree of freedom, v

You may well have appreciated from the PDF that the direct calculation of a χ^2 probability is not a prospect to look forward to. However, as with the normal distribution, we will not be required to perform such calculations, since we can obtain these probabilities from the χ^2 table (see Appendix 3 at the end of the book).

We will now look at several examples whose general theme is as follows: given some data we have collected, how well does it correspond to one of our theoretical distributions? This type of test is called a **goodness-of-fit test** and makes use of the χ^2 distribution.

Fitting a discrete uniform distribution

OCR S3 5.13.5 (a),(b)

Consider the following situation. You have been involved in a game of chance involving gambling. The game involved rolling the usual six-sided die and betting on the outcome (i.e. which of the six numbers showed). You have lost a considerable sum of money and are wondering if the die is actually a fair one. To test this out you have borrowed the die and rolled it 150 times with the following results:

Table 4.1	Outcomes for die
Number showing	*Frequency*
1	23
2	20
3	27
4	26
5	22
6	32
Total	150

The question we would like an answer to is: 'Based on the evidence, is the die a fair one?' The χ^2 test provides us with an answer to the question.

We could argue that we would normally expect an exactly equal distribution given that each number ought to have a one in six chance of showing on each throw. But we also know that such a probability is strictly a long-term average and there remains the question as to whether 150 throws is sufficient. Perhaps if we threw another 150 times the frequencies might begin to balance themselves out more.

It is clear that what we require is some formal method of determining whether the data we have obtained follows an approximately uniform distribution as we would expect. We follow a similar process to the other tests we have examined. Our null hypothesis and alternative hypothesis are:

H_0 : The data follows a uniform distribution.

AH : The data follows a distribution which is not uniform.

Under H_0 we can work out frequencies which we would expect to get and these are shown in Table 4.2.

Table 4.2	Showing observed and expected frequencies	
Number showing	Observed frequency (O)	Expected frequency (E)
1	23	25
2	20	25
3	27	25
4	26	25
5	22	25
6	32	25
Total	150	150

Effectively the null hypothesis is that the observed and expected frequencies are not different.

We now work out the quantity:

$$\sum \frac{(O-E)^2}{E}$$

which is called the χ^2 statistic

It can be shown (but is not required at 'A' level) that:

$$\sum \frac{(O-E)^2}{E} \sim \chi^2 \,(v)$$

where v is the number of **degrees of freedom** of the distribution.

The complete calculation for this example is shown in Table 4.3.

Table 4.3		Calculation of χ^2 for the data			
Number showing	Observed frequency	Expected frequency	$(O-E)$	$(O-E)^2$	$\frac{(O-E)^2}{E}$
1	23	25	−2	4	0.16
2	20	25	−5	25	1
3	27	25	2	4	0.16
4	26	25	1	1	-0.04
5	22	25	−3	9	0.36
6	32	25	7	49	1.96
Total	150	150	0		3.68

So we have $\displaystyle\sum \frac{(O-E)^2}{E} = 3.68$

To find the number of degrees of freedom for this and subsequent tests:

● count up the number of cells (or values that have been compared) which in this case is six

● subtract 1, as in this example only five frequencies are actually independent – if we knew five of the frequencies we could work out the sixth, since the total must be 150

- subtract from this the number of parameters that have been estimated using the data

i.e. $\quad v = k - s - 1$

where $\quad k$ = the number of cells

$\quad\quad\quad s$ = the number of parameters estimated from the data.

In this example we have not used the data to estimate any parameters and so $s = 0$ and $v = k - 1 = 6 - 1 = 5$

Hence our statistic $\sum \dfrac{(O-E)^2}{E} \sim \chi^2(5)$ in this example.

Recall from previous hypothesis testing that we need to fix a level of significance for an hypothesis test. Here, and throughout this section, we will consider only 5% levels and all the tests will be one-tailed, i.e. we reject H_0 if our test statistic (χ^2) lies in the critical region shown in Fig. 4.2.

Figure 4.2

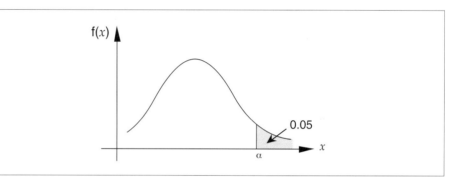

i.e. we will reject H_0 if our calculated value lies to the right of α, as shown. It is precisely the value of α that is given in the χ^2 distribution table in Appendix 3 at the end of the book.

Turning now to the table, we see that in this example where $v = 5$ (v is given in the left-hand column) and with significance level of 0.05 (given in the top row of the table):

- the tabulated value of $\chi^2(5)$ is 11.070 (the critical value) – we will abbreviate this in future to $\chi^2_{5\%}(5)$

- our calculated value was 3.68 – our abbreviation for this is $\chi^2_{calc}(5)$.

Since $\quad \chi^2_{calc}(5) < \chi^2_{5\%}(5)$

we conclude that H_0 should be accepted and that the uniform distribution provides a good model for the data given (at the 5% level of significance).

For tests with other significance levels (e.g. 1%) the procedure is exactly the same except for the value looked up in tables for comparison with the value of $\chi^2_{calc}(5)$, e.g. in this example if we had been conducting a test at the 1% level then we would have compared our value of 3.68 with the tabulated value of 15.086 to be found in the same row but now under the column headed 0.010.

The next example shows the amount of work required for a χ^2 test.

Example	The number of books borrowed from a library during a given week was 135 on Monday, 108 on Tuesday, 120 on Wednesday, 114 on Thursday and 148 on Friday. Is there any evidence at the 5% level of significance that the number of books borrowed depends on the day of the week?

Solution	H_0: The number of books borrowed is independent of the day of the week (i.e. data is uniform).

AH: The number of books borrowed is dependent on the day of the week (i.e. data is non-uniform).

Test at 5% significance level (one-tailed).

Table 4.4

O	E	$O - E$	$(O - E)^2$	$\dfrac{(O - E)^2}{E}$
135	125	10	100	0.800
108	125	−17	289	2.312
120	125	−5	25	0.200
114	125	−11	121	0.968
148	125	23	529	4.232
Total	625	625	0	8.512

Number of degrees of freedom = $5 - 1 = 4$ (no parameters calculated from the data)

$\therefore \qquad \chi^2_{calc}(4) = 8.512$

From the χ^2 distribution table

$$\chi^2_{5\%}(4) = 9.488$$

Since $\chi^2_{calc}(4) < \chi^2_{5\%}(4)$,

we accept H_0.

We conclude that there is no significant difference between the observed data and a uniform distribution, i.e. the number of books borrowed does not depend on the day of the week.

Practice questions A   C 3.2

1   Last year the maths department at a local college achieved 12% A grades, 18% B grades, 40% C grades, 18% D grades and 12% E grades. This year the college achieved 22 A grades, 34 B grades, 66 C grades, 16 D grades and 12 E grades from a class of 150. Is this significant at the 5% level?

2   The number of accidents in a road haulage company over a period of 2 years were tabulated as follows:

Time of day (hours)	$0 \le h < 4$	$4 \le h < 8$	$8 \le h < 12$	$12 \le h < 16$	$16 \le h < 20$	$20 \le h < 24$	Total
Frequency	14	16	24	22	24	20	120

Are these figures consistent with the claim that no one period is more dangerous than any other? (Use 5% significance level.)

3 A die was tossed 60 times with the following results:

Score	1	2	3	4	5	6
Frequency	11	13	9	14	6	7

Test at the 5% level whether the evidence supports the hypothesis that the die is unbiased.

4 The following is taken from a purported random number table.

9674	1148	4426	8908	3945
7019	3392	2224	1321	1388
3541	5321	3551	1864	9197
3015	7285	4351	3284	9788
3148	9189	7128	7199	7121.

Test whether the frequencies of the digits 0, 1, 2, ..., 9 differ from expectation. (Use 5% level of significance.)

Fitting a binomial distribution

OCR S3 5.13.5 (a),(b)

Our next example deals with testing a hypothesis about possible bias in jury selection.

Example

It is known that 55% of potential jurors are female. A random sample of 150 juries gave the distribution shown in Table 4.5.

Does this data approximate to a binomial distribution $X \sim B(150, 0.55)$ or is there some gender bias in the way that juries are selected?

Table 4.5   Selection of female jurors

Number of females on the jury	Number of trials
0	0
1	0
2	2
3	6
4	14
5	25
6	30
7	32
8	24
9	10
10	7
11	0
12	0
Total	150

Solution

We set up two hypotheses:

H_0 : the data comes from the binomial distribution $X \sim B(150, 0.55)$

AH : the data does not come from a binomial distribution

In order to apply the χ^2 test to this data, we first need to calculate the expected frequencies under the null hypothesis that the data is binomial, with $p = 0.55$.

For $X \sim B(150, 0.55)$, $P(X = 0) = (0.45)^{12}$

$P(X = 1) = {}^{12}C_1 (0.45)^{11} (0.55)$

$P(X = 2) = {}^{12}C_2 (0.45)^{10} (0.55)^2$ and so on.

To be able to compare these with the frequencies in Table 4.5, we need to find the corresponding proportions (out of 150) and so each probability must be multiplied by 150.

Completing this process gives Table 4.6.

Table 4.6	Selection of jurors, with expected values	
Number of females on the jury	*O*	*E*
0	0	0.01
1	0	0.15
2	2	1.02
3	6	4.15
4	14	11.43
5	25	22.34
6	30	31.86
7	32	33.38
8	24	25.50
9	10	13.84
10	7	5.08
11	0	1.13
12	0	0.11
Total	150	150.00

At this point it is important to notice that some of the values of E are less than 5 (this didn't occur in the previous examples). In such circumstances we have to combine cells with adjacent ones until each value for E is ≥ 5.

This process is referred to as 'pooling' classes.

This process produces Table 4.7 where the cells for 0, 1, 2 and 3 have been combined and cells for 10, 11 and 12 have been combined. Once this has been done we can then proceed with the test and calculate $\dfrac{(O-E)^2}{E}$ as has been done in the last column.

Table 4.7	Complete jury selection calculations		
Number of females on the jury	*O*	*E*	$\dfrac{(O-E)^2}{E}$
≤ 3	8	5.33	1.34
4	14	11.43	0.58
5	25	22.34	0.32
6	30	31.86	0.11
7	32	33.38	0.06
8	24	25.50	0.06
9	10	13.84	1.07
≥ 10	7	6.22	0.10
Total	150	150.00	3.64

The degrees of freedom for the test are based on the amalgamated table giving

$$v = 8 - 0 - 1 \text{ (since no parameters have been calculated from the data)}$$

i.e. $v = 7$

With $\alpha = 0.05$, we have from the table that $\chi^2_{5\%}(7) = 14.067$ and since $\chi^2_{calc}(7) = 3.64$, we conclude that the null hypothesis should be accepted at the 5% level of significance, i.e. based on the evidence, there is no significant gender bias in selecting juries.

Practice questions B   C 3.2

1   At present 50% of all pigeons are diseased. If 4 pigeons are chosen at random, what is the chance that:

(a) 0   (b) 1   (c) 2   (d) 3   (e) 4

pigeons are diseased?

A collection of 160 sets of 4 pigeons are fed a new antibiotic drug with the result that 19, 49, 50, 36 and 6 of these sets respectively had 0, 1, 2, 3 and 4 diseased. Is the new drug any help in fighting the disease? (Use 1% level of significance.)

2   Pigeons always lay 2 eggs and the chance of an egg hatching is 0.9. What is the chance of 0, 1 and 2 eggs hatching in any clutch of 2?

A new bacteria is applied to 1000 sets of 2 eggs with the result that 50 sets didn't hatch, 160 sets had just one hatch and 790 sets had two hatchings. Does the new bacteria significantly reduce the number of hatchings?

3   The following table shows the number of brown eggs in 1000 boxes which contain 6 eggs:

No. of brown eggs	0	1	2	3	4	5	6
Freq.	97	247	340	217	76	20	3

Estimate the probability that any particular egg is brown, and hence fit a suitable binomial distribution to the data, testing the goodness of fit.

Fitting a Poisson distribution   OCR S3 5.13.5 (a),(b)

The procedure for testing data against some assumed Poisson distribution is similar to that for a binomial. However, we will use this as a means of illustrating another common feature of such examples. With the binomial example we were given the parameters of the assumed distribution. Often we are required to derive them directly from the data itself. (As, for example, in question 3 of the previous Practice questions!)

Example

An engineer has been studying the Poisson distribution, and feels it would be very useful in her current situation where the frequency of breakdown of a particular type of machine is being examined. For the machine type under consideration the following data has been collected:

Table 4.8	Machine breakdowns
Number of breakdowns in a given hour	Frequency
0	50
1	125
2	140
3	105
4	55
5	35
6	17
7	8
Total	535

We pose the question: does this data follow a Poisson process?

Solution

In order to apply the χ^2 test we require expected frequencies derived using Poisson probabilities. But to calculate such probabilities, we require a value for μ, the average value.

Using the standard approach for calculating the mean of a frequency distribution we have:

$$\bar{x} = \frac{\Sigma fx}{\Sigma f} = \frac{1273}{535} = 2.38$$

Using this as the value for μ we can now proceed to determine the Poisson probability for each of the breakdown values, determine the expected frequencies, and then perform the test itself.

The appropriate hypotheses are:

H_0 : The data comes from the Poisson distribution, $X \sim P(2.38)$

AH : The data does not come from a Poisson distribution

Expected number with 0 breakdowns

$$= e^{-2.38} \times 535 \qquad = 49.5$$

Expected number with 1 breakdown

$$= e^{-2.38} \frac{2.38^1}{1!} \times 535 \qquad = 117.8$$

Expected number with 2 breakdowns

$$= e^{-2.38} \frac{2.38^2}{2!} \times 535 \qquad = 140.2$$

Expected number with 3 breakdowns

$$= e^{-2.38} \frac{2.38^3}{3!} \times 535 \qquad = 111.3$$

Similar calculations give:

4 breakdowns = 66.2

5 breakdowns = 31.5

6 breakdowns = 12.5

7 breakdowns = 4.2

If we sum these values, we get 533.2 and the difference arises because the theoretical Poisson distribution which is modelling the data can actually take values greater than 7, albeit with small probabilities. To take account of this, we adjust the final cell so making the totals agree. This produces Table 4.9.

Table 4.9

Number of breakdowns	O	E
0	50	49.5
1	125	117.8
2	140	140.2
3	105	111.3
4	55	66.2
5	35	31.5
6	17	12.5
7 or more	8	6.0
Total	535	535.0

We now check that no pooling of cells is required (there are no E-values less than 5).

The number of degrees of freedom for the test is $8 - 1 - 1 = 6$, since we have used the data to find one parameter and the totals must agree.

$$\chi^2_{calc}(6) = \frac{0.5^2}{49.5} + \frac{7.2^2}{117.8} + \frac{0.2^2}{140.2} + \dots + \frac{4.5^2}{12.5} + \frac{2.0^2}{6.0} = 5.4$$

$\chi^2_{5\%}(6) = 12.59$ at 5% level of significance

and so $\chi^2_{calc}(6) < \chi^2_{5\%}(6)$.

We conclude that at the 5% significance level, there is no evidence to suggest that H_0 should be rejected, i.e. a Poisson distribution with mean 2.38 fits the data.

For practical purposes it is worth noting that an important property of the Poisson distribution (and therefore any data that should be modelled by it) is that the mean and variance are the same. As a preliminary therefore, before conducting the χ^2 test, it is worth calculating values for the sample mean and the sample variance – if these are found to be very different then the Poisson distribution is unlikely to be a useful model for the data.

Practice questions C — C 3.2

1 It is thought that deaths in a hospital follow a Poisson distribution with mean 2 per day.

The number of deaths over the last 102 days was as follows:

Number of deaths	0	1	2	3	4 or more
Frequency	11	30	32	20	9

Does this suggest that the Poisson theory is correct?

2 The number of cars passing a check-point during one hundred 5-minute intervals were recorded as follows:

Number of cars	0	1	2	3	4	5
Frequency	5	23	23	25	14	10

What is the sample mean?

Does a Poisson distribution fit the data?

3 A group of students are required to carry out an experiment in which the results are expected to have a Poisson distribution. The results were:

Score	0	1	2	3	4	5	6	7
Frequency	3	7	12	10	8	5	3	2

How well do the data fit a Poisson distribution?

What conclusion can you draw about the students?

Fitting a normal distribution

OCR S3 5.13.5 (a),(b)

This type of test will be dealt with in detail as there are important differences when the model is continuous and defined over a sample space which is infinite.

Consider the following continuous data which relates to the heights of a sample of a certain variety of shrub 3 weeks after planting.

Height (cm)	0–5	6–10	11–15	16–20	21–25	26–30	More than 31
Number of shrubs	13	21	42	63	38	16	7

The heights are recorded to the nearest centimetre and so for example the interval 6–10 means any height satisfying $5.5 \leq x < 10.5$ would be placed in that interval.

Using these true class boundaries we can calculate the sample mean and sample variance from Table 4.10.

Table 4.10

Height	Frequency	Class boundaries	Mid point	$f \times x$	$f \times x^2$
0–5	13	$0 \leq x \leq 5.5$	2.75	35.75	98.3125
6–10	21	$5.5 \leq x < 10.5$	8	168	1344
11–15	42	$10.5 \leq x < 15.5$	13	546	7098
16–20	63	$15.5 \leq x < 20.5$	18	1134	20412
21–25	38	$20.5 \leq x < 25.5$	23	874	20102
26–30	16	$25.5 \leq x < 30.5$	28	448	12544
More than 31	7	$30.5 \leq x$	33	231	7623
	200			3436.75	69221.3125

(where the mid-point for the final interval was based on interval width the same as the others.)

We get $\bar{x}$ = 17.184 (3 d.p.)

and s^2 = $\left(\dfrac{69\,221.3125}{200} - (17.184)^2 \right) \times \dfrac{200}{199}$

= 51.081 (3 d.p.)

With these preliminaries completed we will now continue with the χ^2 test.

H_0 : The data comes from a normal distribution

AH: The data comes from a distribution which is not normal.

Parameters calculated from the data $\bar{x} = 17.184$ and $s^2 = 51.081$

In order to calculate the expected frequencies under the null hypothesis we now have to find, for the distribution $X \sim N(17.184, 51.081)$ the following probabilities

$P(X < 5.5)$ (since the model extends to $-\infty$)

$P(5.5 \leq X < 10.5)$

$P(10.5 \leq X < 15.5)$ etc.

to $P(X \geq 30.5)$

and it will be these when multiplied up by the factor 200 which will be compared with the observed frequencies.

The calculations are set out in detail below.

Table 4.11

Height	z-value	probability (p)	E(= p × 200)
< 5.5	$z < -1.635$	0.051	10.22
$5.5 \leq x < 10.5$	$-1.635 \leq z < -0.935$	0.124	24.8
$10.5 \leq x < 15.5$	$-0.935 \leq z < -0.236$	0.228	45.6
$15.5 \leq x < 20.5$	$-0.236 \leq z < 0.464$	0.271	54.2
$20.5 \leq x < 25.5$	$0.464 \leq z < 1.164$	0.199	39.8
$25.5 \leq x < 30.5$	$1.164 \leq z < 1.863$	0.091	18.2
$x \geq 30.5$	$z \geq 1.863$	0.0313	6.3
		0.9953	199.12

With practice you will find that these calculations can be completed fairly quickly as the endpoint of one interval is effectively the beginning of another.

To illustrate how the z-numbers are obtained consider the interval $15.5 \leq x < 20.5$.

We are using $X \sim N(17.184, 7.15^2)$ as the model and require $P(15.5 \leq X < 20.5)$.

With the usual standardising procedure this leads to

$$P \left(\frac{15.5 - 17.184}{7.15} \leq z < \frac{20.5 - 17.184}{7.15} \right)$$

$= P(-0.236 \leq z < 0.464)$ rounded to 3 d.p.

Tables are then used to calculate the final probability. The discrepancy in the totals is due to the rounding which was necessary in the calculations and *to make our E-values total exactly to 200, we will adjust the first and last values* by an amount of 0.06 to make the match exact.

Our table for the χ^2 test now is as follows:

Table 4.12

Height	O	E	$\frac{(O-E)^2}{E}$
0–5	13	10.66	0.5137
6–10	21	24.8	0.5823
11–15	42	45.6	0.2842
16–20	63	54.2	1.4288
21–25	38	39.8	0.0814
26–30	16	18.2	0.2659
More than 31	7	6.74	0.0192
Totals	200	200	3.1755

$v = 7 - 2 - 1 = 4$

(since we have 7 cells and 2 parameters were calculated from the data)

Hence $\chi^2_{calc}(4) = 3.2025$

From the χ^2 distribution table, $\chi^2_{5\%}(4) = 9.488$

and since $\chi^2_{calc}(4) < \chi^2_{5\%}(4)$

we conclude that the normal distribution is a good model for the data.

Practice questions D   C 3.2

1 It is thought that the data below follow a normal distribution with mean 10 and variance 100.
(The data was obtained by giving 100 hamsters a new type of feedstuff for a year.)

Gain in mass (g)	Frequency
$x \leq -10$	3
$-10 < x \leq -5$	6
$-5 < x \leq 0$	9
$0 < x \leq 5$	15
$5 < x \leq 10$	24
$10 < x \leq 15$	16
$15 < x \leq 20$	14
$20 < x \leq 25$	8
$25 < x \leq 30$	3
$x > 30$	2

(Use a χ^2 test at the 5% level of significance to test the above hypothesis.

Describe briefly how you would modify this test if the mean and variance were unknown.

2 Use a χ^2 test at the 5% level of significance to determine whether the following data fit a normal distribution.

Height (cm)	150–155	156–161	162–167	168–173	174–179	180–185
Frequency	5	17	38	25	9	6

3 John believes that a certain distribution is normally distributed with mean 10 and standard deviation 1.2. In order to test his theory he chooses a random sample of 200, with the results shown in the table on the right.

x	f
$6.5 \leq x < 7.5$	1
$7.5 \leq x < 8.5$	15
$8.5 \leq x < 9.5$	43
$9.5 \leq x < 10.5$	62
$10.5 \leq x < 11.5$	47
$11.5 \leq x < 12.5$	21
$12.5 \leq x < 13.5$	8
$x > 13.5$	3

He then began to calculate the expected frequencies (according to his hypothesis) but, unfortunately, fell ill before he could complete the task. The table below is as far as he got:

0	1	15	43	62	47	21	8	3
E	3.74	17.38				17.38	3.38	0.36

(a) Complete the E-row for John.

(b) Carry out a χ^2 test at the 5% level of significance and say whether John's theory can be regarded as being correct.

4 Test whether the following sample can be modelled by a normal distribution.

x	$0 \leq x < 2$	$2 \leq x < 4$	$4 \leq x < 6$	$6 \leq x < 8$	$8 \leq x < 10$
f	6	63	192	33	6

Fitting a continuous uniform distribution

OCR S3 5.13.5 (a),(b)

Example

The following data is thought to come from a continuous uniform distribution (where x is measured to the nearest whole number).

x	0–5	6–10	11–15	16–20	21–25	26–30
f	4	12	14	13	6	5

Test this claim at the 5% level.

Solution

If the uniform distribution is appropriate then each frequency should be the same, i.e. we would expect each to be 9 (54 ÷ 6), giving the following table:

O	E	$\frac{(O-E)^2}{E}$
4	9	$\frac{25}{9}$
12	9	$\frac{9}{9}$
14	9	$\frac{25}{9}$
13	9	$\frac{16}{9}$
6	9	$\frac{9}{9}$
5	9	$\frac{16}{9}$
		$\frac{100}{9}$

$v = 6 - 1 = 5$ (since no parameters were required).

$$\chi^2_{calc}(5) = \frac{100}{9} = 11.11$$

From the χ^2 distribution table

$$\chi^2_{5\%}(5) = 11.07$$

and since $\chi^2_{calc}(5) > \chi^2_{5\%}(5)$

we conclude that at a 5% level of significance the null hypothesis should be rejected and the uniform distribution does not provide a good model for the data.

In these circumstances it may be appropriate to consider whether the normal distribution fits the data better. This is set as an exercise at the end of this section (question 7).

Practice questions E C 3.2

1 A sample of values of the continuous variable X were chosen randomly and classified as follows:

x	0–1	1–2	2–3
f	14	33	13

Is there evidence to suggest that $X \sim U(0, 3)$?

2 Test whether the following data fit a continuous uniform distribution:

x	0–2	2–4	4–6	6–8	8–10
f	5	8	7	10	10

3 The weights of a sample of snails were recorded as follows:

Weight (g)	0–10	10–20	20–30	30–40	40–50	50–60
Frequency	8	17	16	23	13	7

Test at the 5% level of significance whether the data fits a continuous uniform distribution.

Contingency tables

OCR S3 5.13.5 (c)

As well as being used to test goodness-of-fit, the χ^2 test can be applied to a particular type of table known as a contingency table.

Frequently, we may be analysing data that can be classed by two (or more) attributes.

Example

Assume that we had collected data on the number of students who fell into each of the pass/fail categories for each sex, male and female.

Table 4.13   **Examination results by sex**

Sex	Pass	Fail	Total
Male	90	30	120
Female	55	25	80
Total	145	55	200

We wish to establish whether a student's sex has had any effect on whether they passed or failed the exam. Effectively we wish to determine whether the two factors are statistically independent of each other, or whether one is contingent (hence the name of the test) upon the other.

Solution

The steps of the test are similar to those of the goodness-of-fit test covered earlier with the exception of the way in which we determine the expected frequencies. Our null hypothesis is that the two factors are independent.

Formally:

H_0 : exam results and gender are independent attributes

AH : exam result is dependent on gender

If this is the case then (from the basic rules of probability) we have:

$$P(Male \cap Pass) = P(Male) \times P(Pass) = \frac{120}{200} \times \frac{145}{200} = 0.435$$

and given that there are 200 students we would expect 87 students that is, 200×0.435, to fall into this particular category, assuming that the null hypothesis is correct.

Continuing in this way:

$$P(Male \cap Fail) = \frac{33}{200} \quad \text{expected frequency} = 33$$
$$P(Female \cap Pass) = \frac{58}{200} \quad \text{expected frequency} = 58$$
$$P(Female \cap Fail) = \frac{22}{200} \quad \text{expected frequency} = 22$$

Note that the totals of the expected frequencies are the same as the actual totals shown in the table.

The number of degrees of freedom for such a test is determined as follows. We have a 2 × 2 table (ignoring the totals) and therefore require four expected values. Given the need to balance rows and columns against the given totals it is clear that once we have calculated one expected value the rest must be determined. This gives one degree of freedom.

We can now proceed to calculate the χ^2 statistic and do the rest of the test.

The separate tables are as follows:

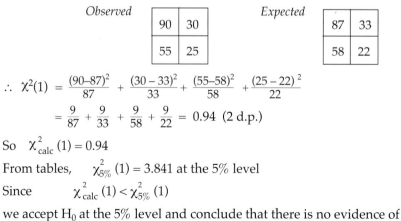

	Observed	
90	30
55	25

	Expected	
87	33
58	22

$$\therefore \ \chi^2(1) = \frac{(90-87)^2}{87} + \frac{(30-33)^2}{33} + \frac{(55-58)^2}{58} + \frac{(25-22)^2}{22}$$

$$= \frac{9}{87} + \frac{9}{33} + \frac{9}{58} + \frac{9}{22} = 0.94 \ (2 \ \text{d.p.})$$

So $\chi^2_{calc}(1) = 0.94$

From tables, $\chi^2_{5\%}(1) = 3.841$ at the 5% level

Since $\chi^2_{calc}(1) < \chi^2_{5\%}(1)$

we accept H_0 at the 5% level and conclude that there is no evidence of dependence between a student's gender and whether they pass or fail.

The test is easily extended to larger contingency tables; the principles remain exactly the same.

> In general the degrees of freedom will be obtained by:
> $$v = (r - 1) \times (c - 1)$$
> where r is the number of rows in the table (excluding totals)
> c is the number of columns (again excluding totals)

Example

A random sample of 100 people were asked by a market research team whether or not they had ever used Sudsey soap. 58 said 'yes' and 42 said 'no'. In a second sample of 80, 62 said 'yes' and 18 said 'no'. A third sample of 75 gave 46 'yes' replies and 29 'no' replies.

Are these three samples showing consistent replies?

Solution

H_0 : The proportion of 'yes' responses is independent of the sample

AH : The proportion of 'yes' responses depends on which sample you choose.

Observed

	1st	2nd	3rd	Total
Yes	58	62	46	166
No	42	18	29	89
Total	100	80	75	255

Expected

	1st	2nd	3rd	Total
Yes	65.1	52.1	48.8	166
No	34.9	27.9	26.2	89
Total	100	80	75	255

(where for example $\dfrac{100 \times 166}{255}$ gives the reading of 65.1 in the expected box and the remaining calculations are done similarly)

$$v = (2 - 1) \times (3 - 1) = 2$$

$$\chi^2_{\text{calc}} (2) = \frac{(58 - 65.1)^2}{65.1} + \frac{(62 - 52.1)^2}{52.1} + \frac{(46 - 48.8)^2}{48.8} + \frac{(42 - 34.9)^2}{34.9} + \frac{(18 - 27.9)^2}{27.9} + \frac{(29 - 26.2)^2}{26.2}$$

$$= 8.073$$

From tables $\chi^2_{5\%} (2) = 5.991$

and we are over this, so we reject H_0.

∴ The proportion of 'yes' responses depends on which sample you choose.

Practice questions F ⚷ C 3.2

1 A random sample produced the following results:

	Deaths due to Cancer	Deaths due to Other causes
Smokers	27	11
Non-smokers	18	44

Are smokers more likely to die from cancer?

2 A random sample produced the following results:

	Alive	Dead
Local anaesthetic	511	24
General anaesthetic	173	21

Does the type of anaesthetic determine your chance of living?

3 For a random sample of 100 rats given diet A, 65 had vitamin deficiency. Another random sample of 100 rats given diet B only had 53 with vitamin deficiency.

Do the proportions of vitamin-deficient rats on the two diets differ significantly?

4 The numbers of female crabs caught on three Norfolk beaches were as follows:

	Beach		
	Sheringham	Cromer	Salthouse
No. of females	44	86	110
Total caught	100	200	200

Is there a significant difference at the 5% level between the number of female crabs on these beaches? (Hint: be careful when setting up your contingency table.)

5 In a survey, 350 car owners from four districts, P, Q, R and S were found to have cars in price ranges A, B, C and D as shown below:

		Districts P	Q	R	S
	A	9	10	12	19
Price	B	13	20	18	29
ranges	C	24	29	12	25
	D	34	41	18	37

Use a χ^2 test to assess the hypothesis that there is no association between the district and the price of the car.

6 Use the following data to test whether there is any significant difference at 5% level in voting pattern between ages.

	Vote Labour	Vote Conservative
21–29	41	28
30–49	95	90
50–65	64	32

SUMMARY EXERCISE

1 A calibrated instrument is used over a wide range of values. To assess the operator's ability to read the instrument accurately, the final digit in each of 700 readings was noted. The results are tabulated below.

Final digit	0	1	2	3	4	5	6	7	8	9
Frequency	75	63	50	58	73	95	96	63	46	81

Use an approximate χ^2 statistic to test whether there is any evidence of bias in the operator's reading of the instrument. Use a 5% significance level and state your null and alternative hypotheses.

2 A market researcher interested in assessing the demand for a new product, advertised the product on local radio for five consecutive days. A telephone number was given so that potential customers could make contact and gain further information.

The results obtained are summarised below.

Day	Mon	Tues	Wed	Thurs	Fri
Number of calls	41	31	44	57	55

Stating clearly your hypotheses and using a 5% level of significance, test whether or not the number of calls received is independent of the day of the week.

3 The following table shows the number of girls in families of 4 children:

Number of girls	0	1	2	3	4
Frequency	15	68	69	38	10

A researcher suggests that a binomial distribution with $n = 4$ and $p = 0.5$ could be a suitable model for the number of girls in a family of 4 children

(a) Test the researcher's suggestion at the 5% level, stating your null and alternative hypotheses clearly.

The research decides to progress to a more refined model and retains the idea of a binomial distribution, but does not specify the value of p, the probability that the child is a girl.

(b) Use the data in the table to estimate p.

The researcher used the value of p in (b) and the refined model, to obtain expected frequencies and found

$$\sum \frac{(O-E)^2}{E} = 2.47 \text{ (There was no pooling of classes.)}$$

(c) Test, at the 5% level, whether the binomial distribution is a suitable model of the number of girls in a family of 4 children.

(d) A family planning clinic has a large number of enquiries from families with 3 boys who would like a fourth child in the hope of having a girl, but they believe their chances are very small. What advice can the researcher give on the basis of the above tests?

4 During hockey practice, each member of a squad of 60 players attempted to hit a ball between two posts. Each player had 8 attempts and the numbers of successes were as follows:

3 4 8 1 0 3 3 4 4 2 6 7 3 2 2 5 5 5 8 1 3 5 6 1 3 4 4 4 1 0
5 3 6 0 6 7 4 3 5 7 0 1 2 6 1 8 0 0 3 0 4 4 1 3 5 0 8 1 8 8

(a) Form the data into an ungrouped frequency distribution.

(b) Use the χ^2 distribution at the 5% significance level to test whether the binomial distribution is an adequate model for the data.

(c) State, giving a reason, whether the data support the view that the probability of success is the same for each player.

5 The number of flaws per 20 metres of a cotton fabric were counted and gave the following frequencies.

Number of flaws	0	1	2	3	4	5	6 or more
frequency	3	7	11	4	5	3	4

Are the data consistent with a Poisson distribution? Test at 5% and 1% significance levels.

6 The following table gives the masses (to the nearest gram) of some items from a factory production line.

Mass (g)	< 995	995–997	998–1000	1001–1004	1005–1007	1008–1010	> 1010
Frequency	2	8	15	32	51	27	18

Is the data consistent with the normal distribution at a 5% significance level?

7 For the data given on p. 54 under the heading 'Fitting a continuous uniform distribution', determine whether the normal distribution provides a better model for the given data.

8 The following table is the result of analysing a random sample of the invoices submitted by branches of a large chain of bookshops.

	Novel	Textbook	General interest
Hardback	24	10	22
Paperback	66	10	18

Using an approximate χ^2 statistic, assess, at the 5% level of significance, whether or not there is any association between the type of book sold and its cover.

State clearly your null and alternative hypotheses.

9 Two schools enter their pupils for a particular public examination and the results obtained are shown below.

	Credit	Pass	Fail
School A	51	10	19
School B	39	10	21

By using an approximate χ^2 statistic, assess at the 5% level of significance, whether or not there is a significant difference between the two schools with respect to the proportions of pupils in the three grades. State your null and alternative hypotheses.

10 (a) Some years ago a Polytechnic decided to require all entrants to a science course to study a non-science subject for one year. In the first year of the scheme entrants were given the choice of studying French or Russian. The number of students of each sex choosing each language is shown in the following table:

	French	Russian
Male	39	16
Female	21	14

Use a χ^2 test at the 5% significance level to test whether choice of language is independent of sex.

(b) The choice of non-science subjects has now been widened and the current figures are as follows:

	French	Poetry	Russian	Sculpture
Male	2	8	15	10
Female	10	17	21	37

Use a χ^2 test at the 5% significance level to test whether choice of subject is independent of sex. In applying the test you should combine French with another subject. Explain why this is necessary and the reasons for your choice.

(c) Point out two features of the data (other than the increase in the number of options and in the total number of students) which have changed markedly over the years.

11 According to Mendel's theory, the obtained proportions should be $9 : 3 : 3 : 1$

    (a) Would observed figures of 315, 108, 101 and 32 cast doubt up on the theory?

    (b) Would observed figures of 137, 44, 43 and 16 cast doubt upon the theory?

SUMMARY   In this section we have seen that:

- the χ^2 (r) statistic is given by $\sum \dfrac{(O-E)^2}{E}$, where O and E represent observed and expected values respectively

- the number of degrees of freedom r is given by

  $r = k - s - 1$

  where k = the number of cells

  and s = the number of parameters estimated from the data

- we reject H_0 in favour of AH if the calculated χ^2 value is larger than the appropriate value in the χ^2 distribution table.

- we amalgamate any E-cells that are less than 5

- the χ^2 test can be used when attempting to fit the following models:

  discrete – uniform, binomial and Poisson

  continuous – normal and uniform

- the χ^2 test can be used on contingency tables where an r by c table has $(r-1)(c-1)$ degrees of freedom.

ANSWERS

Practice questions A

1 $\chi^2_{calc}(4) = 9.785 > \chi^2_{5\%}(4) = 9.488.$

Significant at 5% level. Results better this year.

2 $\chi^2_{calc}(5) = 4.4 < \chi^2_{5\%}(5) = 11.070$

Accept H_0. All periods equally dangerous.

3 $\chi^2_{calc}(5) = 5.2 < \chi^2_{5\%}(5) = 11.07$

Accept H_0. Die is fair.

4 $\chi^2_{calc}(9) = 20.6 > \chi^2_{5\%}(9) = 16.919$

Significant at 5% level. Digits do not occur randomly. (There are too many 1s and too few 0s and 6s.)

Practice questions B

1   (a) $\frac{1}{16}$ (b) $\frac{4}{16}$ (c) $\frac{6}{16}$

     (d) $\frac{4}{16}$ (e) $\frac{1}{16}$

$\chi^2_{calc}(4) = 13.79 > \chi^2_{1\%}(4) = 13.277$

Significant at 1% level.

∴ Drug reduces effects of the disease.

2   0.01, 0.18, 0.81

$\chi^2_{calc}(2) = 162.7.$ Off the table!

Significant at any level. Number of hatchings reduced.

3 $p = \frac{1}{3}$, $\chi^2_{calc}(4) = 4.4$ (1 d.p). Accept H_0.

Binomial model fits with $p = \frac{1}{3}$

Practice questions C

1 $\chi^2_{calc}(4) = 3.77$.

Not significant Accept H_0

Poisson distribution with mean 2 fits.

2 $\bar{x} = 2.5$. $\chi^2_{calc}(4) = 2.59$.

Not significant. Accept H_0

Poisson distribution will mean 2.5 fits

3 $\bar{x} = 3$. $\chi^2_{calc}(3) = 0.28$.

Not significant but *too good to be true*.

It seems possible that the students 'manufactured' their results!

Practice questions D

1 $\chi^2_{calc}(7) = 3.2$

Not significant. Accept H_0.

A normal distribution with mean 10 and variance 100 fits.

If mean and variance were unknown, you would have to estimate them from the sample by finding best estimates. The subsequent number of degrees of freedom would be reduced by two.

2 $\chi^2_{calc}(2) = 2.2$

Not significant. Accept H_0.

A normal distribution with mean 166.54 and variance 51.07 fits

3 (a) Missing cells are 46.56, 64.64, 46.56

(b)

E	21.12	46.56	64.64	46.56	21.12
O	16	43	62	47	32

$\therefore$ $\chi^2_{calc}(4) = 7.23 < \chi^2_{5\%}(4) = 9.488$.
Not significant. Accept H_0.

It seems as if John's theory is correct.

4 $\bar{x} = 4.8$, $s^2 = 1.886$

E	6.21	78.09	158.1	57.6
O	6	63	192	39

$\therefore$ $\chi^2_{calc}(1) = 16.2$

Off the table. A normal distribution does not fit the data.

Practice questions E

1 $\chi^2_{calc}(2) = 12.7$

Off the table. Reject H_0

The uniform distribution does not fit.

2 $\chi^2_{calc}(4) = 2.25$

Not significant. Accept H_0

A continuous uniform distribution fits.

3 $\chi^2_{calc}(5) = 12\frac{6}{7} > \chi^2_{5\%}(5) = 11.070$.
Significant. Reject H_0.

The weight of snails aren't evenly distributed.

Practice questions F

1 E:

17.1	20.9
27.9	34.1

$\therefore$ $\chi^2_{calc}(1) = 16.81$. Off the table! Reject H_0.

There are more cancer deaths from smoking than expected.

2 E:

502	33
182	12

$\therefore$ $\chi^2_{calc}(1) = 9.81$. Off the table. Reject H_0.

More are likely to die with general anaesthetic.

3 E:

59	41
59	41

$\therefore$ $\chi^2_{calc}(1) = 2.98$. Not significant. Accept H_0.

Equal deficiency under both diets.

4 Begin with O:

44	86	110
56	114	90

$\therefore$ $\chi^2_{calc}(2) = 6.6 > \chi^2_{5\%}(2) = 5.991$

Reject H_0. Fewer female crabs than expected at Cromer and more than expected at Salthouse.

5 $\chi^2_{calc}(9) = 12.26$

Not significant. Accept H_0.

There is no association between the district and the price of the car.

6 $\chi^2_{calc}(2) = 6.2 > \chi^2_{5\%}(2) = 5.991$. Reject H_0.

Fewer Labour votes in the age range 30–49 than expected but more than expected in the 50–65 age range.

Regression and correlation

In Section 6 of Unit S1, we investigated linear correlation for bivariate data. In particular we used the test statistic:

$$r = \frac{\sum xy - n\bar{x}\bar{y}}{\sqrt{(\sum x^2 - n\bar{x}^2)(\sum y^2 - n\bar{y}^2)}}$$

for the calculation of the product moment correlation coefficient.

In this section we'll begin by looking at another measure, r_s, of correlation – Spearman's rank correlation coefficient. This will measure the correlation that exists between the orders (or ranks) of the two variables X and Y.

We'll then use our knowledge of hypothesis testing to determine what the vaues of r and r_s tell us about the population from which they were obtained.

Rank correlation

OCR S1 5.11.4 (b),(c)

In some circumstances, it is not necessarily the values of x_i and y_i which may be of interest, but only the relative orders of them.

Example

Suppose that two judges A and B at a flower show award points out of 10 to six exhibits according to the following scheme:

	1	2	3	4	5	6
A	8	3	9	5	0	7
B	9	2	8	7	4	3

Are the judges showing any similarity in their assessments of the merits of the exhibits?

Solution

The actual values of the marks given are not particularly significant here. What is more significant is the order in which they have placed the exhibits.

For this data we will calculate **Spearman's rank correlation coefficient**, a statistic which is simpler to calculate than the correlation coefficient, but which operates on ranks rather than actual data. The formula for the statistic is shown on the following page.

> ### Spearman's rank correlation coefficient
>
> $$r_s = 1 - \frac{6\sum d^2}{n(n^2 - 1)}$$

where d will be explained as we go, and n is the number of pairs of data.

Setting out the relevant table we have

A	B	R_A	R_B	d	d^2
8	9	2	1	1	1
3	2	5	6	−1	1
9	8	1	2	−1	1
5	7	4	3	1	1
0	4	6	4	2	4
7	3	3	5	−2	4
Total				0	12

The column headed d in the table is the difference $R_A - R_B$ where R_A is the ranking of judge A and R_B is the ranking of judge B.

$$r_s = 1 - \frac{6 \times 12}{6 \times 35}$$

$$= 0.657 \text{ (3 d.p.)}$$

The result we have obtained suggests a slight positive correlation between the two judges but that is about all that we can conclude at the moment.

Practice questions A

1 Find Spearman's rank correlation coefficient for the following:

(a)

	Mary	Peter	Jim	Joe	Sally
Maths	23	12	39	54	42
Physics	11	13	12	14	25

(b)

	Alan	Betty	Claire	David
French I	19	70	52	86
French II	17	55	49	41

(c)

	A	B	C	D	E	F
Maths %	72	98	69	71	40	50
English %	71	60	68	69	55	50

2 Two judges independently judge the exhibits of five contestants in a flower show.
The judges rank them as follows:

Contestant	A	B	C	D	E
Judge X	4	3	1	2	5
Judge Y	4	1	2	3	5

Calculate Spearman's rank correlation coefficient for these data. Does your answer confirm that the judges agree reasonably well?

3 Here are the marks of eight candidates in physics and mathematics. Rank the results and hence find Spearman's rank correlation coefficient between the two sets of marks.

Candidate	A	B	C	D	E	F	G	H
Physics mark	43	51	29	82	73	40	38	58
Maths mark	62	65	50	71	32	67	42	49

4 In a beauty competition two judges place the contestants in the following order:

JudgeA   C E D F A G I J B H

Judge B   F G D A I C H E J B

Find Spearman's rank correlation coefficient between the two orders.
[Hint: begin by carefully setting up your table, with the contestants along the top row.]

5

Pupils	A	B	C	D	E	F
Height (cm)	185	177	155	165	160	178
Weight (kg)	70	75	55	50	45	65

Find:

(a) the product moment correlation coefficient

(b) an order for each variable and hence Spearman's rank correlation coefficient.

6 Look at the following scatter diagrams which illustrate four sets of bivariate data A, B, C and D.

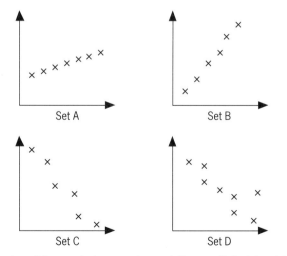

(a) State the value of the product moment correlation coefficient for data set A.

(b) Estimate the value of the product moment correlation coefficient for data set C.

(c) Write down the values of Spearman's rank correlation coefficient for data sets A, B and C.

(d) Why can't you write down the value of Spearman's rank correlation coefficient for data set D?

Ties and Spearman's rank correlation coefficient

To deal with tied ranks in practical work: suppose that a judge ranks the 7th, 8th, 9th and 10th items equally. Then the rank that each should be given is obtained by $\frac{7+8+9+10}{4} = 8.5$, i.e. each item would be ranked as 8.5.

Otherwise proceed as before.

Your syllabus states that 'questions involving ties will not be set but some understanding of how to deal with ties will be expected.'

Example

Consider the table below:

	Alf	Bertha	Cathy	Dan	Edwina
Maths mark	9	9	5	10	8
French mark	4	5	4	9	4

How would you calculate the value of Spearman's rank correlation coefficient?

Solution

In Maths, Alf and Bertha share positions 2 and 3

$\therefore$ They'll each be given positions $\frac{2+3}{2} = 2.5$.

In French, Alf, Cathy and Edwina share positions 3, 4 and 5

$\therefore$ They'll each be given position $\frac{3+4+5}{3} = 4$.

The calculation would then proceed as follows:

	Alf	Bertha	Cathy	Dan	Edwina	
Maths	2.5	2.5	5	1	4	
French	4	2	4	1	4	
d	−1.5	0.5	1	0	0	($\sum d = 0$, as before)
d^2	2.25	0.25	1	0	0	$\sum d^2 = 3.5$

$$\therefore r_s = 1 - \frac{6 \times 3.5}{5 \times 24} = 1 - 0.175 = 0.825.$$

Practice questions B

1 How would you calculate Spearman's rank correlation coefficient for the following?

(a)

	A	B	C	D
Maths I	14	17	18	17
Maths II	5	9	10	10

(b)

	P	Q	R	S	T	U	V
History I	3	8	9	11	8	8	11
History II	4	4	6	10	4	8	4

Hypothesis testing and correlation

To refine our technique of interpreting the value of r or r_s we need to use the technique of hypothesis testing. Essentially, for the purposes of 'A' level, this is an ability to use the tables in Appendix 4 at the end of this book.

No theoretical background is necessary concerning how the tables are obtained but simply the ability to use the tables effectively. We will illustrate their use by considering two examples.

Example

Suppose that we wished to determine whether some specified bivariate data were positively correlated, and a sample of 12 pairs of numbers produced a value of r as 0.53. How should we interpret the value?

Solution

We set up a null hypothesis concerning the population correlation coefficient ρ.

As in previous analyses our calculated value of r gives us an estimate for some unknown value of the population parameter (consisting of all possible pairs!). r is in fact a point estimator of ρ.

Is the value of r obtained significantly greater than zero?
We set up two hypotheses:

$H_0 : \rho = 0$

$AH : \rho > 0$ (i.e. – a one-tailed test)

and choose a significance level (5% being usual).

Referring to the tables and noting the column labelled sample size we now go to the row with 12 in it. Moving to the left we see columns marked 0.10, 0.05, etc.

We are interested in 0.05 (our 5% significance level).

The value given in the tables is 0.4973. This is the critical value. If our calculated value is greater than this value, then we will conclude that there is significant evidence that $\rho > 0$, i.e. that we should reject the null hypothesis.

In our example we found $r = 0.53$. Since $0.53 > 0.4973$ we reject H_0 in favour of AH at the 5% level and so deduce that r is significantly greater than zero, i.e. a positive linear correlation has been demonstrated.

Example

Suppose that we wished to determine whether some specified bivariate data were positively correlated and a sample of 10 pairs of ranked numbers produced a value of r_s as 0.53. How should we interpret this value at the 5% level?

Solution

$H_0 : \rho_s = 0$

$AH : \rho_s > 0$ (one-tailed test)

5% significance level.

Tables (the right-hand columns) give a value of 0.5636 (the critical value)

We have $r_s < 0.5636$

and so we do not reject H_0 in this case. We conclude that there is no significant rank correlation between the variables.

Two-tailed tests

It may be that the question being asked is, 'Are the variables correlated?', without the suggestion or implication of positive or negative. In such cases a two-tailed test should be conducted.

For example, suppose that we wished to test whether there was a correlation between some specified bivariate data and a sample of 12 pairs of items of data produced a value of r as -0.62. The procedure would be as follows:

$$H_0 : \rho = 0$$

$AH : \rho \neq 0$ (two-tailed test)

5% significance level

As we are conducting a two-tailed test the 5% probability has to be split between the two tails giving a probability of 0.025 at each end. This is under the third column of the table and in the case of a sample size of 12 would give the reading 0.576. If our observed correlation coefficient is greater than 0.576 or less than -0.576 then we would reject H_0. In our case $r = -0.62$ is less than the critical value of -0.576 and so we reject H_0 (and accept AH). We conclude that some correlation is shown but without specifying negative or positive.

A caution

It is possible to find sets of data (i.e. sets of pairs) which are highly correlated either positively or negatively, but where it would be incorrect or foolhardy to ascribe a causal connection between the two variables. It could simply be coincidence, or it may be that the two variables are correlated via a third (hidden) variable.

The numerical value of the correlation coefficient can only provide supporting mathematical evidence for a correlation which is suspected from additional evidence or observation. As an illustration it is perfectly plausible that over the last few decades life-expectancy has increased in Britain and that ownership of home computers has also increased. There is no causal connection between these variables – life expectancy has not increased as a result of more people owning home computers and nor has ownership of home computers been affected by life-expectancy. The mathematical correlation between them is related to other factors such as technological advance and increased prosperity.

Practice questions C

1   For each of the following (i) carry out an hypothesis test (ii) state clearly your conclusions.

(a) Are bivariante data *positively* correlated at 5% level?
A sample of 8 readings gave $r = 0.65$.

(b) Are bivariate data *negatively* correlated at 5% level?
A sample of 9 readings gave $r = -0.6$.

(c) Are bivariate data *correlated* at 5% level?
A sample of 10 readings gave $r = 0.6$.

(d) Are bivariate data *correlated* at 5% level?
A sample of 11 readings gave $r = -0.6$.

(e) Are bivariate data *positively* correlated at 5% level?
A sample of 12 *ranked* readings gave $r_s = 0.54$.

(f) Are bivariate data *negatively* correlated at 5% level?
A sample of 13 *ranked* readings gave $r_s = -0.46$.

(g) Are bivariate data *correlated* at 5% level?
A sample of 14 *ranked* readings gave $r_s = 0.52$.

(h) Are bivariate data *correlated* at 5% level?
A sample of 15 *ranked* readings gave $r_s = -0.54$.

2 Applicants for a job with a company are interviewed by two of the personnel staff. After the interviews each applicant is awarded a mark by each of the interviewers. The marks are given below.

	Candidate							
	A	B	C	D	E	F	G	H
Interviewer 1	21	27	24	17	20	22	16	13
Interviewer 2	28	23	25	14	26	17	20	15

(a) Calculate, to 3 decimal places, the Spearman rank correlation coefficient between these two sets of marks.

(b) Assess the statistical significance of your value and interpret your result.

C 3.2

3 The heights h and weights w of 10 people are measured:

$$\sum h = 1710, \quad \sum w = 760, \quad \sum h^2 = 293162, \quad \sum hw = 130628, \quad \sum w^2 = 59390.$$

(a) Calculate, to 4 decimal places, the value of the product moment correlation coefficient.

(b) Stating clearly your hypotheses and using a 5% two-tailed test, interpret this correlation coefficient.

C 3.2

SUMMARY EXERCISE

1 A group of twelve children participated in a psychological study designed to assess the relationship, if any, between age, x years, and average total sleep time (ATST), y minutes. To obtain a measure for ATST, recordings were taken on each child on five consecutive nights and then averaged. The results obtained are shown below.

Child	Age (x year)	ATST (y minutes)
A	4.4	586
B	6.7	565
C	10.5	515
D	9.6	532
E	12.4	478
F	5.5	560
G	11.1	493
H	8.6	533
I	14.0	575
J	10.1	490
K	7.2	530
L	7.9	515

$\sum x = 108 \quad \sum y = 6372 \quad \sum x^2 = 1060.1 \quad \sum y^2 = 3396942 \quad \sum xy = 56825.4$

(a) Calculate the value of the product moment correlation coefficient between x and y. Assess the statistical significance of your value and interpret your results.

(b) Plot these data on a scatter diagram.

Discuss, *briefly*, whether or not your conclusions in (a) should now be amended.

(c) It was subsequently discovered that Child I had been unwell during the study period.

Explain, *without further calculations*, the implications of this additional information on your conclusions.

2 A tasting panel was asked to assess biscuits baked from a new recipe. Each member of the panel was asked to assign a score on a scale from 0 to 100 for texture (X_1), flavour (X_2), sweetness (X_3), chewiness (X_4), and butteriness (X_5).

The scores assigned by the ten members of the panel for texture and flavour were as follows:

Taster	1	2	3	4	5	6	7	8	9	10
X_1	43	59	76	28	53	55	81	49	38	47
X_2	67	82	75	48	91	63	67	51	44	54

(a) Draw a scatter diagram of the data

(b) Calculate the product moment correlation coefficient between X_1 and X_2.

(c) State briefly, how you would expect the scatter diagram in (a) to alter if the tasters were given training in how to assign scores before the tasting took place.

(d) Given that $\sum X_3 = 601$, $\sum X_3^2 = 38637$ and $\sum X_2 X_3 = 40564$ calculate the product moment correlation coefficient between X_2 and X_3.

(e) The table below shows the product moment correlation coefficient between each pair of X_1, X_2, X_3 and X_4 (except for the two calculated in (b) and (d) which have been left blank):

	X_1	X_2	X_3	X_4
X_1	1		0.232	−0.989
X_2		1		−0.478
X_3			1	−0.251
X_4				1

If a decision was made that to save time in future only X_1, X_2 and either X_3 or X_4 would be recorded which variable (X_3 or X_4) would you omit and why?

(f) Given that the correlation coefficient between X_2 and X_5 is exactly 1, what is the correlation coefficient between X_3 and X_5?

Draw up a table showing the numerical value of the product moment correlation coefficient between each pair of X_1, X_2, X_3, X_4 and X_5.

3 The examination marks obtained by *A, B, C, D* and *E* are given as follows:

People	A	B	C	D	E
Mathematics mark	4.6	3.8	4.4	4	3.9
Physics mark	8	1	2	4	5

(a) Work out an order for each subject and calculate Spearman's rank correlation coefficient.

(b) Using the *original data*, work out the product moment correlation coefficient.

4 Six friesian cows were ranked in order of merit at an agricultural show by the official judge and by a student vet. The ranks were as follows:

Official judge	1	2	3	4	5	6
Student vet	1	5	4	2	6	3

(a) Calculate Spearman's rank correlation coefficient between these rankings.

(b) Investigate whether or not there was agreement between the rankings of the judge and the student. State clearly your hypotheses, and carry out an appropriate one-tailed significance test at the 5% level.

5 The data below shows the height above sea level, x metres, and the temperature, $y°C$, at 7.00 a.m., on the same day in summer at 9 places in Europe.

Height (x)	1400	400	280	790	390	590	540	1250	680
Temperature (y)	6	15	18	10	16	14	13	7	13

(a) Plot these data on a scatter diagram.

(b) Calculate the product moment correlation coefficient between x and y.
(Use $\sum x^2 = 5\,639\,200$; $\sum y^2 = 1524$; $\sum xy = 66\,450$.)

(c) Give an interpretation of your coefficient.
On the same day the number of hours of sunshine was recorded and Spearman's rank correlation between hours of sunshine and temperature, based on $\sum d^2 = 28$ was 0.767.

(d) Stating clearly your hypotheses and using a 5% two-tailed test, interpret this rank correlation coefficient.

6 A group of students scored the following marks in their Statistics and Geography examinations.

Student	A	B	C	D	E	F	G	H
Statistics	64	71	49	38	72	55	54	68
Geography	55	50	50	47	65	45	39	82

(a) Find the value of the Spearman rank correlation coefficient between the marks of these students.

(b) Stating your hypotheses and using a 5% level of significance, interpret your value.

SUMMARY   In this section we have:

- used the test statistic $r_s = 1 - \dfrac{6\sum d^2}{n(n^2-1)}$ to calculate **Spearman's rank correlation coefficent**

- seen how to proceed with the calculation of r_s should some positions be tied

- used once again (see Unit S1) the test statistic $r = \dfrac{\sum xy - n\bar{x}\bar{y}}{\sqrt{(\sum x^2 - n\bar{x}^2)(\sum y^2 - n\bar{y}^2)}}$
  for calculating the **product moment correlation coefficient**

- used $H_0 : \rho = 0$
  AH: $\rho > 0$ when looking for **positive** correlation

- used $H_0 : \rho = 0$
  AH: $\rho < 0$ when looking for **negative** correlation

- used $H_0 : \rho = 0$
  AH: $\rho \neq 0$ when looking for **correlation**.

- used the **table of critical values of correlation coefficients** to determine the significance of r and r_s.

ANSWERS

Practice questions A

1 (a)

4	5	3	1	2
5	3	4	2	1

$\sum d^2 = 8$, $r_s = 0.6$

(b) 0.4

(c) $\sum d^2 = 14$, $r_s = 0.60$.

2 $r_s = 0.7$. Close to $+1$ $\therefore$ reasonable agreement.

3 $r_s = 0.14$

4 0.38

5 (a) 0.821

(b) 0.657 ($\sum d^2 = 12$)

6 (a) 1

(b) –0.8 (It must be negative at least.)

(c) 1, 1, –1

(d) As x increases y doesn't decrease $\therefore r_s \neq -1$, it is only near to –1.

Practice questions B

1 (a)

A	B	C	D
4	2.5	1	2.5
4	3	1.5	1.5

(giving $r_s = 0.85$)

(b)

P	Q	R	S	T	U	V
7	5	3	$1\frac{1}{2}$	5	5	$1\frac{1}{2}$
$5\frac{1}{2}$	$5\frac{1}{2}$	3	1	$5\frac{1}{2}$	2	$5\frac{1}{2}$

(giving $r_s = 0.5$)

Practice questions C

1 (a) $H_0 : \rho = 0$

AH : $\rho > 0$

$0.65 > 0.62$. Accept AH.

Positive correlation shown.

(b) $H_0 : \rho = 0$

AH : $\rho < 0$

$0.6 > 0.5822$. Accept AH.

Negative correlation shown.

(c) $H_0 : \rho = 0$

AH : $\rho \neq 0$

$0.5 < 0.6319$. Accept H_0. Uncorrelated.

(d) $H_0 : \rho = 0$

AH : $\rho \neq 0$

$0.6 < 0.6021$. Accept H_0. Uncorrelated.

(e) $H_0 : \rho_s = 0$

AH : $\rho_s > 0$

$0.54 > 0.5035$. Accept AH.

Positive rank correlation shown.

(f) $H_0 : \rho_s = 0$

AH : $\rho_s < 0$

$0.46 < 0.4835$. Accept H_0.

No negative rank correlation shown.

(g) $H_0 : \rho_s = 0$

AH : $\rho_s \neq 0$

$0.52 < 0.5385$. Accept H_0.

No rank correlation shown.

(h) $H_0 : \rho_s = 0$

AH : $\rho_s \neq 0$

$0.54 > 0.5214$. Accept AH.

Rank correlation shown.

2 (a) $\rho_s = 0.452$ ($\sum d^2 = 46$)

(b) Not significant. No correlation demonstrated.

3 (a) 0.6034

(b) $H_0 : \rho = 0$

AH : $\rho \neq 0$

$0.6034 < 0.6319$. Not significant. Accept H_0.

No correlation shown.

[Mind you, if we had been asked to test for positive correlation at the 5% level then, since $0.6034 > 0.5494$, our answer would have been ... yes, positive correlation has been shown. Moral: read the questions carefully!]

Tackling a project

Statistics is a highly practical subject: it has many easily identifiable applications and is an essential part of many areas of study and work. Anyone studying Economics, Politics or a science subject, such as Biology or Physics, will use statistical methods in their work time and again. For example:

- to analyse important economic indicators, such as changes in rates of unemployment or inflation
- to identify trends in voting patterns for different political parties across different regions
- to interpret data gained from a scientific experiment.

Because Statistics is such a practical subject, Edexcel, like most 'A' level Examination Boards, require students to carry out a practical project as part of their 'A' level Statistics. In this section you will investigate what is involved in tackling a Statistics project. This section will also help you to make vital decisions about your project and to work through processes involved in completing a project.

If you are following the OCR specification, much of this section will be of use to you in helping you to carry out a mathematical project for Unit C1. See the OCR syllabus, section 5.17, Mathematics Project.

Carrying out a project will also provide a great many opportunities for gathering evidence of key skills. See p. 74 and Appendix 5 (pp. 98–100) for more details.

C 3.1(a),(b), 3.2, 3.3; **N** 3.1, 3.2, 3.3; **IT** 3.2, 3.3; **WO** 3.2; **LP** 3.1, 3.2, 3.3; **PS** 3.1, 3.2, 3.3

By the end of this section you should be able to:

- identify a subject for a practical project
- understand and follow the procedures involved in tackling a project
- present your project clearly and attractively.

The purpose of your project

Any Statistics project you undertake will have several broad purposes:

- to show your ability to apply statistical techniques in a practical situation
- to demonstrate your mathematical skills in the use of statistical techniques
- to *investigate an area or topic that is of interest to you.*

The Examination Board's syllabus sets out what the overall purpose of your project should be.

The current Edexcel syllabus (pp. 64 and 65) states:

> 'The project should consist of the use of statistical methods to investigate a subject, test an assertion(s) or estimate parameters. To this end, all projects must include data collection...'

> 'The purpose of the project is to demonstrate the ability to apply statistical methods in a practical situation.'

It is important that you check, at the outset, what information the Examination Board supplies in the syllabus you are working to. This information not only includes advice on important factors to bear in mind as you tackle your project, but also gives warnings about things to avoid and traps to avoid falling into.

Size of project

The syllabus gives details of the overall size of the project in terms of recommended length of time to spend on it and, importantly, the number of marks it carries towards your 'A' level.

The current Edexcel syllabus states that the project is worth 25% of the marks awarded for Unit S3 and should represent about 20 hours' work.

Again, check carefully the syllabus you are working to.

A step-by-step approach

Before you begin your project, it will help to have a broad idea of the steps you will need to take. They are outlined below. At each step of the way, you are also likely to have opportunities to use and demonstrate Key skills.

- Plan your project. Choose a project that interests you – it is important that *you* make the choice. **PS** 3.1

- Decide what you want to find out and think how relevant it is to the work you have done. Remember your project should be *practical* and *planned* so that you can demonstrate and use the techniques covered in Units S1, S2 and S3. Don't collect the data first and think of things to do with it afterwards. **LP** 3.1 **PS** 3.1

- Consult with your teacher/tutor for advice regarding your choice of subject, length of project, layout, strategy, etc. **LP** 3.1 **WO** 3.2

- Tackle the project, consulting with your teacher/tutor as you go along. **LP** 3.2, 3.3 **WO** 3.2

- Collect your data – data collection is an essential part of the project. Don't just copy data from secondary sources – you won't have had any influence over the way the data has been presented and cannot be sure of its accuracy or relevance. **N** 3.1

- Analyse the data, representing it in the most suitable form. **N** 3.1, 3.2 **IT** 3.2, 3.3

● Draw conclusions from your analysis.

N 3.3

● Ensure that your project is completed and handed in on a date agreed with your teacher/tutor, or exam centre. Remember: Examination Boards have dates fixed for receiving projects for moderation – if you miss the date because you have not completed your project, you may have to delay taking the relevant examination.

● Review how you carried out the project – what you did well and what improvements you could have made. This will be useful evidence towards Key skills.

LP 3.3
PS 3.3

Structuring your project

Your project will be assessed through the Report that you produce at the end of it. Your Report should be a carefully prepared, clearly structured and well laid out piece of work. Edexcel recommends the following components to your report – other Boards will make their own suggestions:

1 Title

2 Summary

3 Introduction

4 Data collection

5 Analysis of data

6 Interpretation

7 Conclusions

8 Appendix.

As you can see, the structure of the report reflects the different processes you have to go through in tackling the project. We will look at these in more detail.

1 Title

Choose a title which accurately describes your project, so that someone picking it up for the first time has a good idea of what subject you have investigated.

2 Summary

Edexcel recommends a summary between 100 and 200 words describing the main work undertaken and the main conclusions reached. Don't go into any detail here – aim just to give a broad picture.

3 Introduction

This should include:

● a general statement describing what subject you are investigating

● a description of your aims and objectives, including a general statement of a hypothesis or assertion you plan to test

● a description of the methods you plan to use to test the hypothesis or assertion.

Your aims should be phrased in terms of 'to investigate' or 'to test', rather than 'to show', which implies that you have already decided on the outcome of your investigations in advance.

4 Data collection

This should include:

- a description of the method used to collect data
- the reasons for choosing this method
- a description of any problems you encountered in collecting reliable data, e.g. bias, sampling
- where appropriate, a description of why the particular data (rather than other sources of information) were chosen.

There are several possible methods of data collection you could choose:

- a designed experiment
- direct observation
- questionnaires and surveys
- use of secondary data or simulation.

You may be able to use data you have collected for another subject, e.g. Biology, Economics or Chemistry.

At this stage you need to think about what precautions you could take to ensure that the data is free from bias.

- Think carefully about your sample size.
- Watch out for experimental error, e.g. does the time of day or the weather affect your results?
- Be aware of the possible drawbacks of using secondary data.
- Consider carrying out a trial or pilot study to assess the suitability of the method of data collection you have chosen.

When it comes to presenting your Report, avoid putting pages and pages of data in the body of the project. Place the data in an appendix and refer to that.

5 Analysis of the data

The project should include relevant tabular and/or pictorial representations, such as grouped frequency tables, two-way tables, bar charts, pie charts, stem and leaf diagrams, box and wisker plots, scatter graphs and line diagrams. You will have had plenty of practice at constructing these from your work on S1.

The tables and diagrams you include must, however, be relevant to the aims and objectives of the project. Don't create as many different sorts of graph or diagram as you can, just because you can – you will get no credit for that. Select the most relevant representations – explaining why you chose to represent data in a particular way.

Where possible, make any comparisons on the same page.

The project should include the calculation of all relevant statistics, e.g. mean, mode, standard deviation, correlation coefficient. All appropriate workings should be shown. Again, ensure the calculations are *relevant*, not just as many as you can think of.

Where calculations are extensive, include them in an appendix, with just a summary in the body of the report (see point 8 below).

6 Interpretation

Describe or discuss the way in which the data, diagrams and calculations have furthered the aims of your project. (You may find it clearer to include this at relevant points in your analysis of the data – see 5 above.)

7 Conclusions

The project should draw the evidence together in order to state whether or not the hypothesis or assertion made should be rejected. In some circumstances, the project may be inconclusive. This should be stated in a statement of further work which could be undertaken in order to achieve a conclusion.

Criticise your work, e.g. the sample may not have been large enough or it may have been biased because...

Be *precise* when you indicate possible areas for further study.

8 Appendix

This should include copies of extensive calculations, questionnaires, experiment sheets, surveys and raw data.

Examples of approaches to projects

As an example of how to approach a project, skeleton plans for a couple of possible projects are set out below.

Example

Title: An investigation of mail deliveries

Aims:

(a) To investigate whether the daily amount of mail my family receives follows a Poisson distribution.

(b) To determine whether we are more likely to receive mail on any particular day of the week.

(c) To work out what proportion of our mail is junk mail.

Data collection:

Count my family's mail for the next 100 days. Record my results in a table with the following headings.

Date	Day	Ordinary mail	Junk mail	Total mail

(These results would eventually be placed in the appendix.)

At this stage, I might also change course. For example, if the amount of mail received is very small, then it might be better to work in weeks rather than days. On the other hand, perhaps I should be distinguishing between first and second deliveries?

Analysis of data:

In terms of pictorial representation, pie charts would be appropriate.

To answer (a), calculate the sample mean and carry out a χ^2 test. Clearly state my conclusions.

To answer (b), another χ^2 test with H_0: mail equally likely on any day of the week. Clearly state my conclusions.

To answer (c), carry out an hypothesis test following my obtained values for p.

Clearly state my conclusions.

Criticisms of my results:

The 100 days weren't random. Should I have taken a year instead? What about mislaid or mis-delivered mail? No account was taken of whether mail was first or second class. What about parcels?

Possible further studies:

Compare my results with those of one or more neighbours. Is there any correlation between the amount of junk mail received and the time of year?

Example	A biology field trip might lead you investigate the distribution of a particular plant.

Title: An investigation of the growth of A (whatever the plant is).

Aims:

(a) To compare the distribution of numbers of plants on two distinct sites.

(b) To investigate the shape of each plant – stem length v. head diameter.

(c) To investigate the distribution of stem lengths.

Data collection:

On each site, two faces of a ridge, say, divide the area into equal sized squares, number the squares, and choose your sample squares using random numbers.

Count the numbers of plants on these two distinct sites, counting the numbers of plants per square of your grid.

On each site, measure the stem lengths of the plants and their flower diameter (or some other numerical measure, depending on what kind of plant it is).

Analysis of data:

χ^2 contingency table to see if there is a significant difference between the two sites.

Correlation and regression work on stem lengths and flower diameters.

Investigate the distribution of stem lengths – is it normal? χ^2 goodness of fit.

Estimate mean and variance of population – point estimates, confidence intervals.

If other people have done similar work on the same site before (or elsewhere on the same plant), use their data as a base and perform a hypothesis test to see if there is any significant change.

At each stage, produce diagrams as well as calculations – at least they will illustrate your calculations, but they sometimes reveal features which get lost in the number crunching.

Possible further studies:

If no one else's work is available for comparison, then you could plan to repeat all or part of the project yourself, either on the same site or elsewhere.

| **Example** | The use of computer or calculator simulation is encouraged as a means of collecting data. Every calculator has a random number generator, but is it truly random? If it is, then the digits will be uniformly distributed over [0, 9]. |

Title: Random or pseudo-random numbers.

Aims:

(a)   To investigate the randomness of digits produced by a calculator.

(b)   To test the functioning of the Central Limit Theorem.

Note:

Part (a) is testing a null hypothesis that the numbers are U(0, 9).

Part (b) is testing a null hypothesis that the Central Limit theorem correctly describes the increasing normality of the sampling distribution.

Data collection and analysis:

Tally off digits from the calculator and perform a χ^2 goodness of fit of U(0, 9).

Using a programmable calculator, write a program to find the means of random pairs (say) of digits. Tally the results and comment.

Are the mean and variance near what the Central Limit theorem would predict on the basis of a population of U(0, 9)?

Repeat for larger sample sizes (3, 4, etc., and then 40, 50 etc.).

Does the mean stay the same?

Has the variance decreased (as $\dfrac{\sigma^2}{n}$ would predict)?

Is the distribution nearer to being normal (as the Central Limit theorem claims)?

Carry out a goodness of fit test against a normal distribution.

Further work: (some of this might be possible within the scope of this project, depending on how fast you work)

Repeat, using the same calculator on another occasion – does it give the same picture?

Repeat, using a different calculator.

Repeat on a computer, writing a simple programme to make the machine do most of the work for you.

Repeat some of the work on printed random number tables (although beware of the hours of drudgery involved, with less help from machines!).

What project should you do?

Here are a few more topic areas which you could consider as a basis for your project. The ideas are deliberately general – often little more than ideas about data you might collect. It is important that you formulate your own ideas of what to investigate, and how, and that you choose something you are interested in.

Ideas for projects include:

- weather in holiday resorts
- shoe size and glove/hat/dress/shirt size
- colour of cars
- most popular vowel/comparison of different languages
- journey times
- votes in a general or local election
- road accidents
- population investigation/prediction for a particular country
- distributions of bridge hands
- poll on some location questions, e.g. use of public or private transport
- geological section
- the important factors that determine the success of a football team
- plant population
- balance of payment, unemployment or inflation figures
- heights and weights
- effect of temperature on germination rates
- what factors affect the price of a car
- colour test of various chemical solutions
- reaction times for two groups (e.g. male, female) and other psychological experiments
- draws/home wins/away wins – can you fit a model?
- analysis of examination results
- income and expenditure among pupils
- words per page in different sorts of book
- correlation between memory and age
- a person's ability to estimate length.

Tackling a survey

One of the options for data collection mentioned earlier is using a survey or questionnaire. This method has the advantage that you are actively collecting your own data. However, if you choose this approach, you must think carefully about the relevance and purpose of your survey. Assessors see far too many projects based on surveys which have no real relevance or purpose, or are badly conducted or provide doubtful data.

If you do decide to undertake a survey, you will need to think carefully about the design of your questionnaire. Here are some tips to bear in mind:

Designing a questionnaire

- Use simple language.

- Use a small number of questions.

- Be meticulous in how you phrase questions.

- Avoid long, complicated questions.

- Be clear and unambiguous.

- Do not use leading questions.

- Avoid questions outside people's direct experience.

- Avoid embarrassing questions – the respondents may be tempted to mislead.

- Be very careful with lists of alternative answers – people's choices are sometimes based on what they remember or the first or last response in a list.

- Do not rely on people's memories.

- Consider using a pilot study to test out your questionnaire.

Getting a good mark for your project

When marking your project your assessor will be looking at all aspects of the work you have done. In particular, the following areas will be considered:

- the overall subject of the project and the problem being tackled

- the aims and objectives of the project

- the overall strategy in tackling the problem

- method of data collection chosen and effectiveness in collecting the data

- use of relevant and accurate diagrams and calculations

- interpretation of the data collected, in terms of effectiveness, accuracy, relevance, limitations, etc.

- validity of the conclusions drawn

- criticisms made and access for further study discussed.

While it is the content of your Report that you will be assessed on, it is important that you strive to make it as accessible, clear and attractive as possible.

- Your work needs to be direct and compact – don't waffle!

- Write on one side of the paper only. This way you are less likely to make mistakes and it is easier for the assessor to mark.

- Write up your work as you go along. This will save you time and will also ensure that you don't forget, or lose, anything.

- If you make a mistake, cross it out neatly and carry on.

- Consider using a word processor to present your report.

- You can also use a computer for diagrams and calculations. If you do this, it is important to show in your report that you understand the purpose and implications of the diagrams and calculations.

SUMMARY

This section has focused on what is involved in writing an 'A' level Statistics project. It contains a great deal of advice about the structure, content and presentation of your project. Re-read the guidance given here as many times as you need to. Above all remember the following important guidelines:

- Your project needs to be practical and relevant, with a clear purpose.

- Try to choose a project that gives you the maximum opportunity to demonstrate a wide range of skills.

- Try to choose a project that you are genuinely interested in.

- Above all, discuss your project with your teacher or tutor – make the most of their advice.

$S3$

Practice examination paper

(Attempt all 8 questions.)

1   Independent random variables X and Y are such that: $E(X) = 50$, $Var(X) = 3$, $E(Y) = 42$ and $Var(Y) = 4$.

   Find the mean and variance of the following:

   (a) $X - Y$

   (b) $2X + Y - 142$.

   Which of the above answers (if any) would be affected if X and Y were no longer independent?

2   What is meant by a 'random sample'? Briefly describe a situation in which a quota sample might be more appropriate.

   The following is a sequence of random digits:

   9015 9964 6596 8583 7357 7247 1304
   9811 3209

   Use them in order to select the following random samples, making your method clear:

   (a) two months from the 12 in a year

   (b) a point in the right-angled triangle with sides of length 5, 5, $5\sqrt{2}$ cm.

3   A sack contains apples whose masses have a normal distribution with mean 75.2g and standard deviation 1.4g. Calculate the probability that the mean mass of a sample of 16 randomly chosen apples will exceed 74.5g.

4   Three judges in a music competition place the six finalists in the following order:

Competitor	A	B	C	D	E	F
First judge	5	1	4	2	6	3
Second judge	6	2	3	5	4	1
Third judge	6	3	1	4	5	2

   Find Spearman's rank correlation coefficient between the first two judges and also between the first and third judges. Do the findings of the first judge agree better with those of the second or third judge?

5   A product moment correlation coefficient based on a sample of size 27 was computed to be 0.4. Can we conclude at a significance level of:

   (a) 5%   (b) 1%,

   that the corresponding population coefficient differs from zero?

6   Of a group of patients who complained that they did not sleep well, some were given sleeping pills while others were given a placebo. They were later asked whether the pills helped them or not. Their responses are given below:

	Slept well	Didn't sleep well
Took sleeping pills	44	10
Took a placebo	81	35

   Assuming that all patients told the truth, test the hypothesis that there is no difference between sleeping pills and the placebo at the 5% significance level.

7   A sample of 100 light bulbs produced by manufacturer A showed a mean lifetime of 1200 hours and a standard deviation of 90 hours. A sample of 75 light bulbs produced by manufacturer B showed a mean of 1240 hours with a standard deviation of 120 hours. Making any necessary assumptions, test the hypothesis that the bulbs of manufacturer B are superior to those of manufacturer A at the 1% level.

   Explain why your assumptions were necessary.

continues on p. 84

8 A machine is supposed to produce keys to a nominal length of 5.00 cm. A random sample of 50 keys produced by the machine was such that:

$$\sum x = 250.50 \text{ cm} \quad \text{and} \quad \sum x^2 = 1255.0290 \text{ cm}^2,$$

where x denotes the length of a randomly chosen key produced by the machine.

(a) Calculate unbiased estimates of the mean and variance of the length of keys produced by the machine.

(b) Explain what you understand by 'unbiased estimate'.

(c) Calculate a 90% symmetric confidence interval for the mean length of keys produced by the machine.

(d) If an α% symmetric confidence interval for the mean length of keys produced by the machine has a width of 0.01227 cm, what is the value of α?

(e) Carry out an hypothesis test at the 5% level to determine whether the machine is producing keys which are too long (i.e. greater than 5.00 cm). State clearly the null and alternative hypotheses as well as your conclusion.

$S3$

Solutions

Section 1

1 (a) $E(5X + 7)$ $= 5E(X) + 7$
                           $= 5 \times 20 + 7 = 107$

(b) $Var(5X + 7)$ $= Var(5X)$
                           $= 5^2 Var(X) = 25 \times 2 = 50$

(c) $E(5X + 7Y)$ $= 5E(X) + 7E(Y)$
                           $= 5 \times 20 + 7 \times 24 = 268$

(d) $Var(5X + 7Y)$ $= 5^2 Var(X) + 7^2 Var(Y)$
                           $= 5^2 \times 2 + 7^2 \times 3 = 197$

(e) $E(5X - 7Y)$ $= 5E(X) - 7E(Y)$
                           $= 5 \times 20 - 7 \times 24 = -68$

(f) $Var(5X - 7Y)$ $= 5^2 Var(X) + 7^2 Var(Y)$
                           $= 5^2 \times 2 + 7^2 \times 3 = 197$
                           (Note this + sign ↑)

(g) $E(5 - 7Y)$ $= 5 - 7E(Y)$
                           $= 5 - 7 \times 24 = -163$

(h) $Var(5 - 7Y)$ $= Var(-7Y)$
                           $= 7^2 Var(Y) = 7^2 \times 3 = 147$

2 (a) $p(1) = k$
     $p(2) = 8k$
     $p(3) = 27k$
     $p(4) = 64k$
     $k + 8k + 27k + 64k = 100k = 1$

     $\Rightarrow k = \dfrac{1}{100}$

(b) $E(R)$ $= 1 \times \dfrac{1}{100} + 2 \times \dfrac{8}{100} + 3 \times \dfrac{27}{100}$
                 $+ 4 \times \dfrac{64}{100}$

            $= \dfrac{354}{100} = 3.54$

$E(R^2)$ $= 1^2 \times \dfrac{1}{100} + 2^2 \times \dfrac{8}{100}$
            $+ 3^2 \times \dfrac{27}{100} + 4^2 \times \dfrac{64}{100}$

           $= \dfrac{1300}{100} = 13$

          $\Rightarrow Var(R) = 13 - 3.54^2 = 0.4684$

(c) $E(5R - 3) = 5E(R) - 3 = 14.7$
     $Var(5R - 3) = 25Var(R) = 11.71$

3 (a) $3X$ has distribution

r	0	3	6
$P(3X = r)$	$\dfrac{1}{3}$	$\dfrac{1}{2}$	$\dfrac{1}{6}$

(b) $2Y$ has distribution

r	-2	2
$P(2Y = r)$	$\dfrac{3}{4}$	$\dfrac{1}{4}$

(c) $3X + 2Y$ has the distribution

r	-2	1	2	4	5	8
$P(3X + 2Y = r)$	$\dfrac{6}{24}$	$\dfrac{9}{24}$	$\dfrac{2}{24}$	$\dfrac{3}{24}$	$\dfrac{3}{24}$	$\dfrac{1}{24}$

(d) $E(X) = 0 \times \dfrac{1}{3} + 1 \times \dfrac{1}{2} + 2 \times \dfrac{1}{6} = \dfrac{5}{6}$

     $E(X^2) = 0 \times \dfrac{1}{3} + 1^2 \times \dfrac{1}{2} + 2^2 \times \dfrac{1}{6} = \dfrac{7}{6}$

     $\Rightarrow Var(X) = \dfrac{7}{6} - \left(\dfrac{5}{6}\right)^2 = \dfrac{17}{36}$

(e) $E(Y) = (-1) \times \dfrac{3}{4} + 1 \times \dfrac{1}{4} = \dfrac{-1}{2}$

     $E(Y^2) = 1 \times \dfrac{3}{4} + 1 \times \dfrac{1}{4} = 1$

     $Var(Y) = 1 - \left(\dfrac{-1}{2}\right)^2 = \dfrac{3}{4}$

(f) $E(3X + 2Y) = \dfrac{36}{24}$

     $E\left((3X + 2Y)^2\right) = \dfrac{228}{24}$

     $Var(3X + 2Y) = \dfrac{29}{4}$

4 (a)

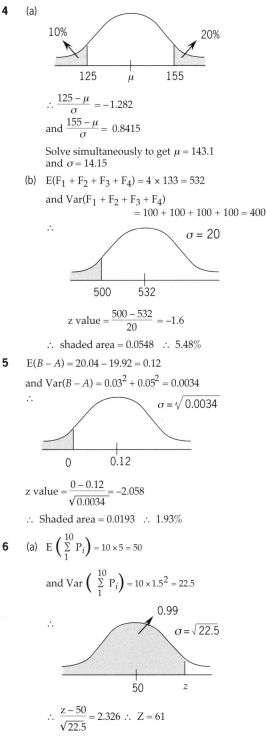

$$\therefore \frac{125 - \mu}{\sigma} = -1.282$$

and $\frac{155 - \mu}{\sigma} = 0.8415$

Solve simultaneously to get $\mu = 143.1$
and $\sigma = 14.15$

(b) $E(F_1 + F_2 + F_3 + F_4) = 4 \times 133 = 532$

and $Var(F_1 + F_2 + F_3 + F_4)$
$$= 100 + 100 + 100 + 100 = 400$$

$\therefore$

$\sigma = 20$

z value $= \dfrac{500 - 532}{20} = -1.6$

$\therefore$ shaded area $= 0.0548$ $\therefore$ 5.48%

5 $E(B - A) = 20.04 - 19.92 = 0.12$

and $Var(B - A) = 0.03^2 + 0.05^2 = 0.0034$

$\therefore$

$\sigma = \sqrt{0.0034}$

z value $= \dfrac{0 - 0.12}{\sqrt{0.0034}} = -2.058$

$\therefore$ Shaded area $= 0.0193$ $\therefore$ 1.93%

6 (a) $E\left(\sum_1^{10} P_i\right) = 10 \times 5 = 50$

and $Var\left(\sum_1^{10} P_i\right) = 10 \times 1.5^2 = 22.5$

$\therefore$

0.99

$\sigma = \sqrt{22.5}$

$\therefore \dfrac{z - 50}{\sqrt{22.5}} = 2.326$ $\therefore$ $Z = 61$

$\therefore$ 11.01

(b) This time we have $\sum_1^{22} P_i$

$\therefore$

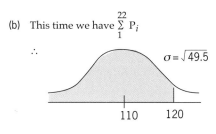

$\sigma = \sqrt{49.5}$

$\sigma = \sqrt{49.5}$ $\therefore$ shaded area $= 0.9222$

Section 2

1 (a) (i) There are many solutions, such as selecting every 10th person entering a supermarket for a survey, or testing every 1000th light bulb from a production line.

   (ii) Stratified sampling is generally used when the population divides naturally into sub-groups and it is desirable to sample in proportion to the numbers in the sub-groups.

(b) One major advantage of stratified sampling is that of arriving at a more representative sample where there are natural sub-groups in the population.

   A disadvantage is that the sample obtained is no longer truly random.

2 Stratified

It is important to gain information from each of the departments. Stratified sampling would ensure each department is represented in the sample. The Principal might find some variation between departments, e.g. science students might be happier generally to forego sport in favour of extra lectures. This might help decide which departments should have lectures on Wednesdays and which should not.

3 Criticisms are:

(a) Not everybody has a phone and some people are ex-directory.

(b) A member of a large household has less chance of being chosen than a member of a small household.

(c) This excludes under 18's, Peers and those who failed to register.

(d) This excludes those that are housebound, away on holiday or working.

Probably (a) or (c) would give the best representative sample.

Section 3

1 90% interval $= 10.66 \pm 1.65 \cdot \dfrac{10.76}{\sqrt{2000}}$

$\qquad\qquad\qquad = 10.66 \pm 0.397$

$\qquad\qquad\qquad = (10.26, 11.06)$

2 99% interval $= 266 \pm 2.58 \times \dfrac{20}{\sqrt{40}}$

$\qquad\qquad\qquad = 266 \pm 8.16$

$\qquad\qquad\qquad = (257.84, 274.16)$

3 $\bar{x} = 1000.48$

$\quad s^2 = 3.52$

(a) 95% interval $= 1000.48 \pm 1.96 \times \dfrac{1.88}{\sqrt{10}}$

$\qquad\qquad\qquad\qquad = 1000.48 \pm 1.16$

$\qquad\qquad\qquad\qquad = (999.32, 1001.64)$

(b) 99% interval $= 1000.48 \pm 2.5758 \times \dfrac{1.88}{\sqrt{10}}$

$\qquad\qquad\qquad\qquad = 1000.48 \pm 1.53$

$\qquad\qquad\qquad\qquad = (998.95, 1002.01)$

(c) Require $1.96 \times \dfrac{1.88}{\sqrt{n}} < 0.6$

$\qquad \Rightarrow n > 37.7$

$\qquad \Rightarrow n = 38$

4 (a) $\bar{X} \sim N\left(\mu, \dfrac{\sigma^2}{n}\right)$

(b) Require $1.96 \times \dfrac{40}{\sqrt{n}} < 15$

$\qquad \Rightarrow n > 27.3$

$\qquad \Rightarrow n$ at least 28

5 $\bar{x} = 499.27$ (2 d.p.)

(a) 90% interval is $499.27 \pm 1.645 \times \dfrac{7.5}{\sqrt{11}}$

$\qquad = (495.6, 503.0)$

(b) $H_0 : \mu = 502$

$\qquad$ AH $: \mu < 502$

$\qquad$ one-tailed test

$\qquad 5\%$ significance level

$\qquad z = \dfrac{499.27 - 502}{\dfrac{7.5}{\sqrt{11}}} = -1.207$

$\qquad z > -1.645 \Rightarrow$ Accept H_0.

(c) If $\mu = 496$ then z is positive and > -1.645 obviously. So no need for this one-tailed test.

6 $E(\bar{X}) = 50$ and $Var(\bar{X}) = \dfrac{4}{N}$

$\therefore \dfrac{50.2 - 50}{\sqrt{\dfrac{4}{N}}} = 2$ (Look up 0.9772 in the tables.)

$\therefore 0.2 = 2\sqrt{\dfrac{4}{N}}$

$\therefore \sqrt{N} = 2\dfrac{\sqrt{4}}{0.2}$ $\quad \therefore N = 400$ exactly.

7 Best estimate of (public) population S.D

$= 19\sqrt{\dfrac{200}{199}} = 19.05$

Best estimate of (private) population S.D

$= 16\sqrt{\dfrac{300}{299}} = 16.03$

$H_0: \mu_1 = \mu_2 \quad$ AH$: \mu_1 < \mu_2$

$\therefore \overline{\text{Pub}} - \overline{\text{Pri}} \approx N\left(10, \dfrac{19.05^2}{200} + \dfrac{16.03^2}{300}\right),$

i.e. $\approx N(0, 2.67)$.

$\therefore$ P (difference ≥ 6) has Z value 3.67. Off the table!

Accept AH.
Public sector pay less than private sector pay.

Section 4

1 $\chi^2_{\text{calc}} = 38.2$

$\quad \nu = 9$

$\quad \chi^2_{5\%} = 16.92$

$\quad$ Reject H_0 – evidence of bias

2 $\chi^2_{\text{calc}} = 9.98$

$\quad \nu = 4$

$\quad \chi^2_{5\%} = 9.49$

$\quad$ Reject H_0 – number of calls not independent of the day of the week.

3 (a) Expected frequencies are

$\qquad 12.5 \quad 50 \quad 75 \quad 50 \quad 12.5$

$\qquad H_0$: Data is binomial

$\qquad$ AH : Data has some other distribution

$\qquad 5\%$ significance

$\qquad \chi^2_{\text{calc}} = 0.5 + 6.48 + 0.48 + 2.88 + 0.5$

$\qquad\qquad\quad = 10.84$

$\qquad \nu = 4$

$\qquad \chi^2_{5\%} = 9.488$

$\qquad$ Reject H_0

(b) $\bar{x} = 1.8 \Rightarrow 4p = 1.8 \Rightarrow p = 0.45$

(c) $\nu = 3$

$\qquad \chi^2_{5\%} = 7.815$

$\qquad H_0$: Data is binomial

$\qquad$ AH : Data is not binomial

$\qquad$ Since $2.47 < 7.815$, accept H_0

(d) From tests in (a) and (c), the better model for the number of girls in a family of 4 is $X \sim B(4, 0.45)$

$\qquad$ The chance of a girl is therefore just under $\dfrac{1}{2}$ at each trial.

4 (a)

x	f		$p(x)$	E	
0	8		0.0084	0.504	
1	8	20	0.0548	3.288	group 13.206
2	4		0.1569	9.414	
3	10		0.2568	15.408	
4	9		0.2627	15.762	
5	7		0.1719	10.314	
6	5		0.0703	4.218	
7	3	14	0.0164	0.984	group 5.304
8	6		0.0017	0.102	
	60				

(b) $p = 0.45$

H_0 : Binomial distribution B(8, 0.45) models the data

$v = 3$ (5 cells, 1 parameter estimated
from the data and totals agree)

$\chi^2_{calc} = 23.6, \quad \chi^2_{5\%} = 7.815$

Reject H_0

(c) Since binomial is not suitable model,
unlikely that p *is* constant.

5 $\bar{x} = \dfrac{100}{37} = 2.703$

H_0 : $X \sim P(2.703)$

AH : X is NOT $\sim P(2.703)$

x	0	1	2	3	4	5	6 +
E	2.48	6.70	9.07	8.16	5.51	2.98	1.15

x = Number of flaws per 20 m

E = Expected frequencies (2 d.p.)

Combine the cells for $x = 0$ and $x = 1$

Combine cells $x = 4$, $x = 5$ and $x = 6+$

x	≤ 1	2	3	≥ 4
E	9.18	9.06	8.16	9.64

This leads to $\chi^2_{calc} = 3.18$

Now $v = 4 - 1 - 1 = 2$ (since $\bar{x}$ calculated from data)

$\chi^2_{5\%}(2) = 5.991, \quad \chi^2_{1\%}(2) = 9.21$

Hence accept H_0 at both significance levels

6 $\sum f = 153$

$\sum x_i = 153\,784$

$\sum x_i^2 = 154\,575\,056$

(based on mid-points 993, 996, 999, 1002.5, 1006, 1009, 1012)

$\bar{x} = 1005.124183$

$s^2 = \dfrac{154\,575\,056}{153} - (1.005.124)^2 = 20.2283$

But, rather than aproximate the value of the meaning when calculating s^2, we may write

$s^2 = \dfrac{154\,575\,056}{153} - \left(\dfrac{153\,784}{153}\right)^2 = 19.8604$

Using this last calculation, an unbiased estimate for $\text{Var}(x)$ is

$19.8604 \times \dfrac{153}{152} = 19.99$, so that $\sigma = 4.47$.

This leads to the following table, based on
H_0 : $x \sim N(1005.124, 19.99)$

and hence AH: x is not distributed as N(1005.124, 19.99)

$a \leq x < b$	$Z = \dfrac{\text{upper class boundary} - 1005.124}{4.47}$	$\Phi(z)$	$P(a \leq X < b)$	E	Combined O	Combined E	χ^2
$-\infty < x < 994.5$	-2.377	0.0088	0.0088	1.3464			
$994.5 \leq x < 997.5$	-1.706	0.0440	0.0352	5.3856	10	6.7320	1.5864
$997.5 \leq x < 1000.5$	-1.034	0.1425	0.0985	15.0705	15	15.0705	0.0003
$1000.5 \leq x < 1004.5$	-0.140	0.4443	0.3018	46.1754	32	46.1754	4.3517
$1004.5 \leq x < 1007.5$	0.532	0.7026	0.2583	39.5199	51	39.5199	3.3348
$1007.5 \leq x < 1010.5$	1.203	0.8855	0.1829	27.9837	27	27.9837	0.0346
$1010.5 \leq x < \infty$	∞	1.0000	0.1145	17.5185	18	17.5185	0.0132
			1.0000	153.0	153		9.3210

Hence $\chi^2_{calc} = 9.3210$

Now $v = 6 - 2 - 1 = 3$

$\chi^2_{5\%}(3) = 7.815$

Hence, since $9.321 > 7.815$, we reject H_0 and conclude that the Normal distribution is NOT a good fit at the 5% level.

7 $\sum x_i = 801$

$\sum x_i^2 = 14\,470.25$

$\sum f = 54$

$\Rightarrow \quad \bar{x} = 14.833$

$\quad s^2 = 48.844$

H_0 : Data has normal distribution

AH : Data has some other distribution

$a < x \le b$	$Z = \dfrac{b - 14.833}{6.989}$	$P(a < X \le b)$	E
$-\infty < x < 5.5$	-1.34	0.090	4.865
$5.5 \le x < 10.5$	-0.62	0.179	9.671
$10.5 \le x < 15.5$	0.10	0.272	14.699
$15.5 \le x < 20.5$	0.81	0.251	13.565
$20.5 \le x < 25.5$	1.53	0.146	7.884
$25.5 \le x < \infty$	∞	0.062	3.316
		1.000	54

The first two and the last two classes must be pooled. This gives the following table.

O	E
16	14.537
14	14.699
13	13.565
11	11.200

$\chi^2_{calc} = 0.208$

$\nu = 4 - 2 - 1 = 1$

$\chi^2_{5\%} (1) = 3.841$

Accept the hypothesis that the normal distribution is a good fit at the 5% level.

8 $\left.\begin{array}{l} H_0 : \text{no association} \\ AH : \text{association} \end{array}\right\}$ between type and cover

Expected frequencies

	N	T	G
H	33.6	7.47	14.93
P	56.4	12.53	25.07

$\nu = (3-1)(2-1) = 2$

$\chi^2_{calc} = 11.09$

$\chi^2_{5\%} (2) = 5.99$

Reject H_0

There is an association.

9 H_0 (independence)

Expected frequencies are

	C	P	F
A	48	10.7	21.3
B	42	9.3	18.7

$\nu = 2$

$\chi^2_{calc} = 0.1875 + 0.0458 + 0.2484 + 0.2143$
$\qquad + 0.0527 + 0.2829 = 1.032$

$\chi^2_{5\%} (2) = 5.99.$ Accept H_0

10 (a) Expected frequencies as follows

	F	R
μ	36.7	18.3
F	23.3	11.7

$\chi^2_{calc} = \dfrac{(39 - 36.7)^2}{36.7} + \dfrac{(16 - 18.3)^2}{18.3} + \dfrac{(21 - 23.3)^2}{23.3} + \dfrac{(14 - 11.7)^2}{11.7}$

$\qquad = 1.112$

$\chi^2_{5\%} (1) = 3.841.$ Accept H_0.

(b) Observed:

	F/R	P	S	
M	17	8	10	35
F	31	17	37	85
	48	25	47	120

Expected:

14	7.3	13.7
34	17.7	33.3

$\chi^2_{calc} = 0.643 + 0.067 + 0.999 + 0.265 + 0.028 + 0.411$

$\qquad = 2.413$

$\chi^2_{5\%} (2) = 5.991$

Accept H_0 (independence)

French / Russian combined as observed French < 5 and these are known to be independent.

(c) French decreased in popularity. Female Russian students increased in spite of alternatives.

11 (a)

O	315	108	101	32
E	312.75	104.25	104.25	34.75

$\therefore \chi^2_{calc} (3) = 0.47.$ Not significant. However, the χ^2 value calculated here is so small that the observed figures could have been 'manufactured'. Re-trial suggested.

(b)

O	137	44	43	16
E	135	45	45	15

$\therefore \chi^2_{calc} (3) = 0.207.$ Not significant. However, the χ^2 value is too small for comfort. It seems likely that the person who collected the data 'fixed' his findings so that the model was justified. Re-trial suggested with another researcher!

Section 5

1 (a) $r = -0.481$

$H_0 : \rho = 0$

$AH : \rho < 0$ 5% significance

Critical value = −0.4973

Accept H_0 ∴ No significant correlation shown.

(b) See scatter diagram below.

Conclusion should be amended. There is a trend with *I* as serious outlier.

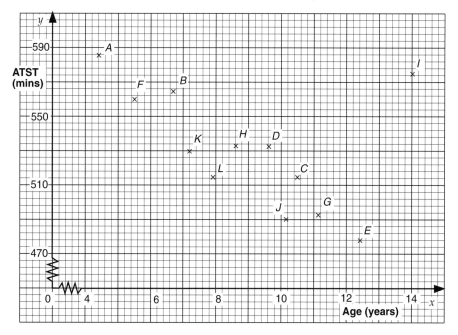

(c) Child *I* needed more sleep owing to illness

2 (a) See scatter diagram below.

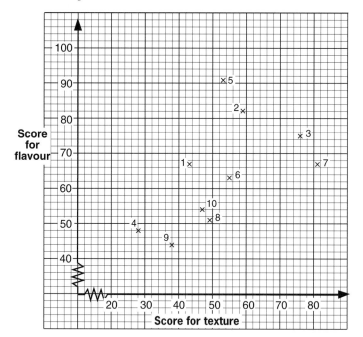

(b) $r_2 = 0.549$

(c) become closer to straight line

(d) $r_3 = 0.858$

(e) Omit X_4 – 'chewiness' is highly negatively correlated with the others.

(f) 0.858

	X_1	X_2	X_3	X_4	X_5
X_1	1	0.549	0.232	–0.989	0.549
X_2		1	0.858	–0.478	1
X_3			1	–0.251	0.858
X_4				1	–0.478
X_5					1

3 (a) The subject orders are as follows:

	A	B	C	D	E
Maths	1	5	2	3	4
Physics	1	5	4	3	2

∴ d 0   0   –2   0   2

∴ $\Sigma d^2 = 0 + 0 + (-2)^2 + 0 + 2^2 = 8$

∴ $r_s = 1 - \dfrac{6 \times 8}{5(5^2 - 1)} = 0.6$

(b) $\bar{M} = 4.14$,

Var(M) = 0.0944

$\bar{P} = 4$

Var(P) = 6, $\Sigma MP = 84.9$, Covariance = 0.42

∴ $r = \dfrac{0.42}{\sqrt{0.0944 \times 6}} = 0.56$

4 (a) 0.314

(b) $H_0 : \rho_s = 0$

AH : $\rho_s > 0$

Accept H_0 – no agreement

5 (a) The scatter diagram is shown below.

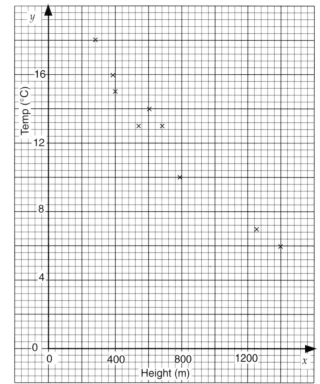

Height (m)

(b) -0.98

(c) Strong negative correlation suggests as height increases temperature decreases

(d) $H_0 : \rho_s = 0$

AH : $\rho_s > 0$

Reject H_0 – association exists.

6 (a) 0.637

(b) $H_0 : \rho_s = 0$

AH : $\rho_s > 0$

Accept H_0 ∴ no correlation shown.

Appendix 1: Random numbers

86 13	84 10	07 30	39 05	97 96	88 07	37 26	04 89	13 48	19 20
60 78	48 12	99 47	09 46	91 33	17 21	03 94	79 00	08 50	40 16
78 48	03 37	82 26	01 06	64 65	94 41	17 26	74 66	61 93	24 97
80 56	90 79	66 94	18 40	97 79	93 20	41 51	25 04	20 71	76 04
99 09	39 25	66 31	70 56	30 15	52 17	87 55	31 11	10 68	98 23
56 32	32 72	91 65	97 36	56 61	12 79	95 17	57 16	53 58	96 36
66 02	49 93	97 44	99 15	56 86	80 57	11 78	40 23	58 40	86 14
31 77	53 94	05 93	56 14	71 23	60 46	05 33	23 72	93 10	81 23
98 79	72 43	14 76	54 77	66 29	84 09	88 56	75 86	41 67	04 42
50 97	92 15	10 01	57 01	87 33	73 17	70 18	40 21	24 20	66 62
90 51	94 50	12 48	88 95	09 34	09 30	22 27	25 56	40 76	01 59
31 99	52 24	13 43	27 88	11 39	41 65	00 84	13 06	31 79	74 97
22 96	23 34	46 12	67 11	48 06	99 24	14 83	78 37	65 73	39 47
06 84	55 41	27 06	74 59	14 29	20 14	45 75	31 16	05 41	22 96
08 64	89 30	25 25	71 35	33 31	04 56	12 67	03 74	07 16	49 32
86 87	62 43	15 11	76 49	79 13	78 80	93 89	09 57	07 14	40 74
94 44	97 13	77 04	35 02	12 76	60 91	93 40	81 06	85 85	72 84
63 25	55 14	66 47	99 90	02 90	83 43	16 01	19 69	11 78	87 16
11 22	83 98	15 21	18 57	53 42	91 91	26 52	89 13	86 00	47 61
01 70	10 83	94 71	13 67	11 12	36 54	53 32	90 43	79 01	95 15

Appendix 2: The normal distribution function

The function tabulated below is $\Phi(z)$, defined as $\Phi(z) = \dfrac{1}{\sqrt{2\pi}} \displaystyle\int_{-\infty}^{z} e^{-\frac{1}{2}t^2} \, dt$.

z	$\Phi(z)$	z	$\Phi(z)$	z	$\Phi(z)$	z	$\Phi(z)$	z	$\Phi(z)$
0.00	0.5000	0.50	0.6915	1.00	0.8413	1.50	0.9332	2.00	0.9772
0.01	0.5040	0.51	0.6950	1.01	0.8438	1.51	0.9345	2.02	0.9783
0.02	0.5080	0.52	0.6985	1.02	0.8461	1.52	0.9357	2.04	0.9793
0.03	0.5120	0.53	0.7019	1.03	0.8485	1.53	0.9370	2.06	0.9803
0.04	0.5160	0.54	0.7054	1.04	0.8508	1.54	0.9382	2.08	0.9812
0.05	0.5199	0.55	0.7088	1.05	0.8531	1.55	0.9394	2.10	0.9821
0.06	0.5239	0.56	0.7123	1.06	0.8554	1.56	0.9406	2.12	0.9830
0.07	0.5279	0.57	0.7157	1.07	0.8577	1.57	0.9418	2.14	0.9838
0.08	0.5319	0.58	0.7190	1.08	0.8599	1.58	0.9429	2.16	0.9846
0.09	0.5359	0.59	0.7224	1.09	0.8621	1.59	0.9441	2.18	0.9854
0.10	0.5398	0.60	0.7257	1.10	0.8643	1.60	0.9452	2.20	0.9861
0.11	0.5438	0.61	0.7291	1.11	0.8665	1.61	0.9463	2.22	0.9868
0.12	0.5478	0.62	0.7324	1.12	0.8686	1.62	0.9474	2.24	0.9875
0.13	0.5517	0.63	0.7357	1.13	0.8708	1.63	0.9484	2.26	0.9881
0.14	0.5557	0.64	0.7389	1.14	0.8729	1.64	0.9495	2.28	0.9887
0.15	0.5596	0.65	0.7422	1.15	0.8749	1.65	0.9505	2.30	0.9893
0.16	0.5636	0.66	0.7454	1.16	0.8770	1.66	0.9515	2.32	0.9898
0.17	0.5675	0.67	0.7486	1.17	0.8790	1.67	0.9525	2.34	0.9904
0.18	0.5714	0.68	0.7517	1.18	0.8810	1.68	0.9535	2.36	0.9909
0.19	0.5753	0.69	0.7549	1.19	0.8830	1.69	0.9545	2.38	0.9913
0.20	0.5793	0.70	0.7580	1.20	0.8849	1.70	0.9554	2.40	0.9918
0.21	0.5832	0.71	0.7611	1.21	0.8869	1.71	0.9564	2.42	0.9922
0.22	0.5871	0.72	0.7642	1.22	0.8888	1.72	0.9573	2.44	0.9927
0.23	0.5910	0.73	0.7673	1.23	0.8907	1.73	0.9582	2.46	0.9931
0.24	0.5948	0.74	0.7704	1.24	0.8925	1.74	0.9591	2.48	0.9934
0.25	0.5987	0.75	0.7734	1.25	0.8944	1.75	0.9599	2.50	0.9938
0.26	0.6026	0.76	0.7764	1.26	0.8962	1.76	0.9608	2.55	0.9946
0.27	0.6064	0.77	0.7794	1.27	0.8980	1.77	0.9616	2.60	0.9953
0.28	0.6103	0.78	0.7823	1.28	0.8997	1.78	0.9625	2.65	0.9960
0.29	0.6141	0.79	0.7852	1.29	0.9015	1.79	0.9633	2.70	0.9965
0.30	0.6179	0.80	0.7881	1.30	0.9032	1.80	0.9641	2.75	0.9970
0.31	0.6217	0.81	0.7910	1.31	0.9049	1.81	0.9649	2.80	0.9974
0.32	0.6255	0.82	0.7939	1.32	0.9066	1.82	0.9656	2.85	0.9978
0.33	0.6293	0.83	0.7967	1.33	0.9082	1.83	0.9664	2.90	0.9981
0.34	0.6331	0.84	0.7995	1.34	0.9099	1.84	0.9671	2.95	0.9984
0.35	0.6368	0.85	0.8023	1.35	0.9115	1.85	0.9678	3.00	0.9987
0.36	0.6406	0.86	0.8051	1.36	0.9131	1.86	0.9686	3.05	0.9989
0.37	0.6443	0.87	0.8078	1.37	0.9147	1.87	0.9693	3.10	0.9990
0.38	0.6480	0.88	0.8106	1.38	0.9162	1.88	0.9699	3.15	0.9992
0.39	0.6517	0.89	0.8133	1.39	0.9177	1.89	0.9706	3.20	0.9993
0.40	0.6554	0.90	0.8159	1.40	0.9192	1.90	0.9713	3.25	0.9994
0.41	0.6591	0.91	0.8186	1.41	0.9207	1.91	0.9719	3.30	0.9995
0.42	0.6628	0.92	0.8212	1.42	0.9222	1.92	0.9726	3.35	0.9996
0.43	0.6664	0.93	0.8238	1.43	0.9236	1.93	0.9732	3.40	0.9997
0.44	0.6700	0.94	0.8264	1.44	0.9251	1.94	0.9738	3.50	0.9998
0.45	0.6736	0.95	0.8289	1.45	0.9265	1.95	0.9744	3.60	0.9998
0.46	0.6772	0.96	0.8315	1.46	0.9279	1.96	0.9750	3.70	0.9999
0.47	0.6808	0.97	0.8340	1.47	0.9292	1.97	0.9756	3.80	0.9999
0.48	0.6844	0.98	0.8365	1.48	0.9306	1.98	0.9761	3.90	1.0000
0.49	0.6879	0.99	0.8389	1.49	0.9319	1.99	0.9767	4.00	1.0000
0.50	0.6915	1.00	0.8413	1.50	0.9332	2.00	0.9772		

Percentage points of the normal distribution

The values z in the table are those which a random variable $Z \sim N(0, 1)$ exceeds with probability p; that is, $P(Z > z) = 1 - \Phi(z) = p$.

p	z	p	z
0.5000	0.0000	0.05000	1.6449
0.4000	0.2533	0.0250	1.9600
0.3000	0.5244	0.0100	2.3263
0.2000	0.8416	0.0050	2.5758
0.1500	1.0364	0.0010	3.0902
0.1000	1.2816	0.0005	3.2905

Appendix 3: Percentage points of the χ^2 distribution

The values in the table are those which a random variable with the χ^2 distribution on ν degrees of freedom exceeds with the probability shown.

ν	0.995	0.990	0.975	0.950	0.900	0.100	0.050	0.025	0.010	0.005
1	0.000	0.000	0.001	0.004	0.016	2.705	3.841	5.024	6.635	7.879
2	0.010	0.020	0.051	0.103	0.211	4.605	5.991	7.378	9.210	10.597
3	0.072	0.115	0.216	0.352	0.584	6.251	7.815	9.348	11.345	12.838
4	0.207	0.297	0.484	0.711	1.064	7.779	9.488	11.143	13.277	14.860
5	0.412	0.554	0.831	1.145	1.610	9.236	11.070	12.832	15.086	16.750
6	0.676	0.872	1.237	1.635	2.204	10.645	12.592	14.449	16.812	18.548
7	0.989	1.239	1.690	2.167	2.833	12.017	14.067	16.013	18.475	20.278
8	1.344	1.646	2.180	2.733	3.490	13.362	15.507	17.535	20.090	21.955
9	1.735	2.088	2.700	3.325	4.168	14.684	16.919	19.023	21.666	23.589
10	2.156	2.558	3.247	3.940	4.865	15.987	18.307	20.483	23.209	25.188
11	2.603	3.053	3.816	4.575	5.580	17.275	19.675	21.920	24.725	26.757
12	3.074	3.571	4.404	5.226	6.304	18.549	21.026	23.337	26.217	28.300
13	3.565	4.107	5.009	5.892	7.042	19.812	22.362	24.736	27.688	29.819
14	4.075	4.660	5.629	6.571	7.790	21.064	23.685	26.119	29.141	31.319
15	4.601	5.229	6.262	7.261	8.547	22.307	24.996	27.488	30.578	32.801
16	5.142	5.812	6.908	7.962	9.312	23.542	26.296	28.845	32.000	34.267
17	5.697	6.408	7.564	8.672	10.085	24.769	27.587	30.191	33.409	35.718
18	6.265	7.015	8.231	9.390	10.865	25.989	28.869	31.526	34.805	37.156
19	6.844	7.633	8.907	10.117	11.651	27.204	30.144	32.852	36.191	38.582
20	7.434	8.260	9.591	10.851	12.443	28.412	31.410	34.170	37.566	39.997
21	8.034	8.897	10.283	11.591	13.240	29.615	32.671	35.479	38.932	41.401
22	8.643	9.542	10.982	12.338	14.042	30.813	33.924	36.781	40.289	42.796
23	9.260	10.196	11.689	13.091	14.848	32.007	35.172	38.076	41.638	44.181
24	9.886	10.856	12.401	13.848	15.659	33.196	36.415	39.364	42.980	45.558
25	10.520	11.524	13.120	14.611	16.473	34.382	37.652	40.646	44.314	46.928
26	11.160	12.198	13.844	15.379	17.292	35.563	38.885	41.923	45.642	48.290
27	11.808	12.879	14.573	16.151	18.114	36.741	40.113	43.194	46.963	49.645
28	12.461	13.565	15.308	16.928	18.939	37.916	41.337	44.461	48.278	50.993
29	13.121	14.256	16.047	17.708	19.768	39.088	42.557	45.722	49.588	52.336
30	13.787	14.953	16.791	18.493	20.599	40.256	43.773	46.979	50.892	53.672

Appendix 4: Critical values for correlation coefficients

These tables concern tests of the hypothesis that a population correlation coefficient ρ is 0.
The values in the tables are the minimum values which need to be reached by a sample correlation coefficient in order to be significant at the level shown, on a one-tailed test.

Product Moment Coefficient					Sample	Spearman's Coefficient		
		Level			size		Level	
0.10	0.05	0.025	0.01	0.005		0.05	0.025	0.01
0.8000	0.9000	0.9500	0.9800	0.9900	4	1.0000	–	–
0.6870	0.8054	0.8783	0.9343	0.9587	5	0.9000	1.0000	1.0000
0.6084	0.7293	0.8114	0.8822	0.9172	6	0.8286	0.8857	0.9429
0.5509	0.6694	0.7545	0.8329	0.8745	7	0.7143	0.7857	0.8929
0.5067	0.6215	0.7067	0.7887	0.8343	8	0.6429	0.7381	0.8333
0.4716	0.5822	0.6664	0.7498	0.7977	9	0.6000	0.7000	0.7833
0.4428	0.5494	0.6319	0.7155	0.7646	10	0.5636	0.6485	0.7455
0.4187	0.5214	0.6021	0.6851	0.7348	11	0.5364	0.6182	0.7091
0.3981	0.4973	0.5760	0.6581	0.7079	12	0.5035	0.5874	0.6783
0.3802	0.4762	0.5529	0.6339	0.6835	13	0.4835	0.5604	0.6484
0.3646	0.4575	0.5324	0.6120	0.6614	14	0.4637	0.5385	0.6264
0.3507	0.4409	0.5140	0.5923	0.6411	15	0.4464	0.5214	0.6036
0.3383	0.4259	0.4973	0.5742	0.6226	16	0.4294	0.5029	0.5824
0.3271	0.4124	0.4821	0.5577	0.6055	17	0.4142	0.4877	0.5662
0.3170	0.4000	0.4683	0.5425	0.5897	18	0.4014	0.4716	0.5501
0.3077	0.3887	0.4555	0.5285	0.5751	19	0.3912	0.4596	0.5351
0.2992	0.3783	0.4438	0.5155	0.5614	20	0.3805	0.4466	0.5218
0.2914	0.3687	0.4329	0.5034	0.5487	21	0.3701	0.4364	0.5091
0.2841	0.3598	0.4227	0.4921	0.5368	22	0.3608	0.4252	0.4975
0.2774	0.3515	0.4133	0.4815	0.5256	23	0.3528	0.4160	0.4862
0.2711	0.3438	0.4044	0.4716	0.5151	24	0.3443	0.4070	0.4757
0.2653	0.3365	0.3961	0.4622	0.5052	25	0.3369	0.3977	0.4662
0.2598	0.3297	0.3882	0.4534	0.4958	26	0.3306	0.3901	0.4571
0.2546	0.3233	0.3809	0.4451	0.4869	27	0.3242	0.3828	0.4487
0.2497	0.3172	0.3739	0.4372	0.4785	28	0.3180	0.3755	0.4401
0.2451	0.3115	0.3673	0.4297	0.4705	29	0.3118	0.3685	0.4325
0.2407	0.3061	0.3610	0.4226	0.4629	30	0.3063	0.3624	0.4251
0.2070	0.2638	0.3120	0.3665	0.4026	40	0.2640	0.3128	0.3681
0.1843	0.2353	0.2787	0.3281	0.3610	50	0.2353	0.2791	0.3293
0.1678	0.2144	0.2542	0.2997	0.3301	60	0.2144	0.2545	0.3005
0.1550	0.1982	0.2352	0.2776	0.3060	70	0.1982	0.2354	0.2782
0.1448	0.1852	0.2199	0.2597	0.2864	80	0.1852	0.2201	0.2602
0.1364	0.1745	0.2072	0.2449	0.2702	90	0.1745	0.2074	0.2453
0.1292	0.1654	0.1966	0.2324	0.2565	100	0.1654	0.1967	0.2327

Appendix 5: Key Skills

Your work on this book will provide plenty of opportunities for gathering evidence towards Key Skills, especially in Communication, but also in all the other Key skills. Carrying out a project (see Section 6) will be especially useful in this respect.

These opportunities are indicated by the 'key' icon, e.g. ⚷ **C** 3.2. This means that the exercise contains the type of task that is relevant to Communication Level 3 and may help you gather evidence specifically for C3.2.

The places where Key Skills references are given are listed below, together with some other ideas about possible opportunities for gathering evidence.

Communication

C 3.1a   Contribute to a group discussion about a complex subject.

C 3.1b   Make a presentation about a complex subject, using at least one image to illustrate complex points.

When you have completed your project, you could present your findings to a group and then take part in a question and answer session. This would cover both C3.1a and C3.1b.

C 3.2   Read and synthesise information from two extended documents that deal with a complex subject. One of these documents should include at least one image.

See pages 15, 17, 18, 26, 33, 35, 45, 48, 50, 53, 55, 58 and 69 of this book. See also Section 6.

Your work on S3 provides plenty of opportunities for reading and synthesising information, especially if you carry out a project that involves analysing tables or statistical diagrams.

C 3.3   Write two different types of documents about complex subjects. One piece of writing should be an extended document and include at least one image.

Your project report, if it is well written and follows the guidelines in Section 6, should fully meet the requirement for an extended document about a complex subject.

Application of number

N 3.1   Plan, and interpret information from two different types of sources, including a large data set.

If the topic that you choose for your project involves handling large amounts of data, then it should be capable of generating evidence relating to N 3.1.

N 3.2   Carry out multi-stage calculations to do with:

    (a)   amounts and sizes;

    (b)   scales and proportion;

    (c)   handling statistics;

    (d)   rearranging and using formulae.

You should work with a large data set on at least **one** occasion.

N 3.3   Interpret results of your calculations, present your findings and justify your methods. You must use at least one graph, one chart and one diagram.

As with communication, writing up your project and presenting a report on it should provide lots of scope for presenting results effectively, and for discussing and interpreting results.

Information technology

IT 3.2   Explore, develop, and exchange information and derive new information to meet two different purposes.

In preparing your project report, you may have opportunities to use IT for numerical and/or graphical work.

IT 3.3   Present information from different sources for two different purposes and audiences. Your work must include at least one example of text, one example of images and one example of numbers.

You will have plenty of opportunities to use IT to help in the preparation and presentation of your project report.

Working with others

WO 3.2   Seek to establish and maintain co-operative working relationships over an extended period of time, agreeing changes to achieve agreed objectives.

Although your project needs to be your own individual work for assessment purposes, you may have opportunities to gather evidence for WO 3.2, e.g. in discussing, agreeing and reviewing objectives with a tutor or mentor.

Improving own learning and performance

LP 3.1   Agree targets and plan how these will be met over an extended period of time, using support from appropriate people.

The skills listed for LP 3.1 will be essential for planning your statistics project. For example, you will need to check that the project is suitable for assessment purposes, to agree a target date for submitting your report of a written project and to plan how to organise the time in order to meet this target.

LP 3.2   Take responsibility for your learning by using your plan, and seeking feedback and support from relevant sources, to help meet targets.

As you work on your project, you will need to implement a plan, change methods of approach as necessary and discuss progress with your teacher or tutor at agreed intervals.

LP3.3   Review progress on **two** occasions and establish evidence of achievements, including how you have used learning from other tasks to meet new demands.

During the course of your project you could show evidence of how you achieved your objectives and could then discuss with your teacher or tutor what action you will take in the future to improve your performance.

Problem-solving

PS 3.1   Explore a complex problem, come up with **three** options for solving it and justify the option selected for taking further.

PS 3.2   Plan and implement at least **one** option for solving the problem, review progress and revise your approach as necessary.

PS 3.3   Apply agreed methods to check if the problem has been solved, describe the results and review your approach to problem solving.

In planning and working on your project, you could demonstrate how you explored different options to approach your task and could explain how you took steps to break the overall task down into manageable steps and to plan and work methodically.

Once you have completed your project, you could communicate, both verbally and in writing, what approaches you took to solving problems and what conclusions you reached.